1980

aspects
of
biophysics

aspects of biophysics

William Hughes
Bowdoin College

John Wiley & Sons
New York   Chichester
Brisbane   Toronto

Text and Cover designed by Mark E. Safran
Production was supervised by Linda R. Indig
Copy editing was supervised by Ellen MacElree

Library of Congress Cataloging in Publication Data:

Hughes, William, 1936—
    Aspects of biophysics.

    Bibliography: p.
    Includes index.
    1. Biological physics.   I. Title.
QH505.H78   1979   574.1'91   78-8992
ISBN 0-471-01990-9

Printed in the United States of America

10 9 8 7 6 5 4 3 2 1

To AGMH and TAH

preface

This book is intended for those who have a basic knowledge of physics, chemistry, and biology and who wish to see how some of the more elementary parts of physics may be applied to the study of living matter. The treatment of the applications given here is, of course, not encyclopedic; it could not and was not intended to be. Instead, some representative topics from biophysics are presented, along with references that can guide the student to more detailed and thorough discussions of these topics.

Biophysics is a vigorous and still growing subject. In many cases, it would be easy to carry the development of certain topics into regions of far too uncertain knowledge. I have tried to avoid this by focusing on general results, methods, and analyses, so that the material is not quite so likely to be made useless by the natural progress of the subject.

I have profited from discussions with many, but especially with the late Dr. Peter Curran, who first introduced me to biophysics through his lectures on irreversible thermodynamics. Of course, those who have assisted me bear no responsibility for any errors in these pages.

William Hughes

to the student

In a recent National Academy of Sciences study, it was concluded that "the classical subdisciplines of biology are insufficiently instructive as approaches to current understanding and appreciation of life in its varigated manifestations." Thus classical zoology, botany, and microbiology, for example, are not used as subdivisions in the academy report. Instead the bases for living phenomena are examined at increasingly higher levels of organization, beginning with molecules and proceeding to organelles; cells; tissues and organs; organisms; and finally to species and ecosystems. To a considerable extent, the organization of this book, though not its scope, reflects the above view. Some comments on how the chapter topics fit into the above view of biology can now be made.

The first two chapters discuss the most basic ways of learning about the physical aspects of biological macromolecules, as opposed to their chemical properties, and the small-scale structure of biological material. The study of ultrastructure, that is, the description of how living matter is seen to be put together when studied at resolutions ~ 4 to $\sim 10^2$ nm, is a field all its own, marked by the use of many empirical techniques for preparing the material and extensive qualitative discussion of the observations. Discussions of much of this very detailed structure have been left out for two reasons: to keep the text from becoming too long and because the physical analysis does not yet carry into this realm, especially not at the mathematical level used here. Of course, the progress of the subject must eventually lead to a detailed analysis that will involve the use of such detailed structural knowledge.

The second chapter discusses several techniques for studying macromolecules and subcellular components. Again, the approach is general and does not aim to introduce any substantial amounts of detailed results. This is not because such results are not important but only because the level of the text does not require them. The methods discussed are in general use in both biophysics and biochemistry, but the precise details of any particular method will depend on its application. For example, just how one prepares a centrifuge tube depends on what one wants to separate, and how one adjusts a spectrophotometer depends on what one wants to study and what model of spectrophotometer is at hand. Clearly, these details are best left for discussion in the context of actual laboratory work.

Chapter 3 introduces some of the most elementary ideas for describing the behavior and properties of macromolecules. Two important classes of biological macromolecules are then discussed in Chapters 4 and 5. Again, much information that would be important in a purely biological context but that is not needed at this level of physics has been omitted; for example, there is no detailed discussion of a variety of specific enzyme reactions.

The discussion of the cell membrane in Chapter 6 follows my established trend in that it focuses on particular aspects of the membrane that are both important and can be treated at an appropriate level. In this case I am concentrating on transmembrane phenomena such as transport, and the many other critical biological phenomena in which the membrane plays a major role, such as the immune response, are not discussed. Of course, the study of membranes is now such an immense subject that this chapter can only serve to introduce the topic. In a similar way, all of the complications and detailed knowledge of neurons are not discussed, but instead a central problem, the conduction in the axon, receives the most attention in Chapter 7. Again, the biological details of neuron structure, function, and organization have outrun the ability of an analysis at this level to make much of a contribution. Finally, in Chapter 8, an introduction to the most fundamental aspects of artificial membranes is given, but the specific laboratory details of preparation and a discussion of the use of such membranes to investigate specific biological phenomena is best given in the context of actual laboratory problems.

Chapter 9 is concerned with energy transduction, but of necessity there can be no discussion of the very successful efforts in biochemistry that have produced detailed knowledge about the chemical reactions of metabolism, nor is there any more than the most basic remarks about the structures involved in these processes.

In Chapter 10 the basic ideas involved in understanding radiation effects are presented. Of course, there is a great deal of information on the radiation responses of particular organisms, as well as for particular tissues and organs, which does not appear here. In addition, information about the therapeutic uses of radiation has been omitted and should be sought in the appropriate medical volumes and journals.

Chapters 11 to 15 deal with physiological matters and it is fair to say that each gives some analysis of a topic that is of central importance to the subject. One should be aware, however, that a very large amount of information on the detailed structure and physiological responses of the relevant organs and tissues has been accumulated and can be found in the appropriate physiology books and journals. The structures shown here give only the basic information needed for the physical analysis.

The problem of the origin of life is one that is far less speculative and far more promising as a "proper" subject than was once the case. Nevertheless, the mathematical demands are impressive, especially in model building, and a substantial knowledge of biochemistry is probably going to be required. Recognizing these limits, in Chapter 16, I introduce the problem in a generally qualitative way and point the way to the quantitative treatments.

The last chapter shows some areas where technology based on biophysics is already important. Bioengineering is already an independent subject and this chapter can only serve to introduce it to the reader.

W.H.

contents

aspects
of
biophysics

introduction

"Life is a little current of electricity, driven by sunlight." No one would deny that the study of electricity and light are proper topics of physics, and thus Szent-Gyorgyi's remark is all the justification needed for the study of biological materials by physicists. Of course, there are more significant justifications. For one, an understanding of the intricate processes and complex structural arrangements of living matter seems unlikely to be achieved except through insights that only physics and chemistry can provide. The contributions to those insights that come from physics make up the subject that we call, for convenience, biophysics.

Although it is scarcely a new part of science, the precise origin of biophysics is a matter of debate. The importance of investigating biological phenomena was realized by physicists such as Helmholtz (1821–1894), who, because of his outstanding studies of the physics of hearing and vision, is often cited as the founder of biophysics. These early applications of physics to physiology are both well-known and important, but it should also be noted that cellular processes were not neglected; for example, the electrical properties of cells were the subject of fundamental studies in the late 1800s. However, these first investigations also represent relatively isolated efforts; biophysics during this time was waiting upon the growth of physics, the development of the capacity to investigate, both experimentally and theoretically, increasingly complex states of matter.

In a certain sense, physics has only recently reached a state at which matter as complexly organized as that in biological systems can be studied with understanding. The interest in applying modern physics to such systems certainly was strongly stimulated by Erwin Schrodinger's famous essay, 'What is Life?," in which he argued that the study of these complex systems, in addition to being intrinsically interesting, might also profit physics through the discovery of new physical laws. Although this prediction has not yet proved to be true, the recent successes of physicists and chemists in unraveling biological puzzles has provided ample evidence of the intriguing mechanisms and behavior exhibited by living matter. Of course, biophysics knows no shortage of real problems with as yet unknown solutions, nor is there a guarantee that the solutions of some outstanding problems do not depend explicitly on further developments. Schrodinger may yet be proved correct: the complexity of biological systems may provide the stimulus for the development of intriguing new pieces of physics whose applications will range over both biological and nonbiological problems.

What are the fundamental properties exhibited by living matter? We can either take a very simple view, and thus be able to write down some general answers, or we can consider the many complexities and finally end bogged down in an attempt to "define life." Seymor Benzer once observed that scientists could be divided into "clarifiers" (obviously complimentary) and "turbidifiers" (obviously not) depending, respectively, on their willingness to simplify complex problems. Indeed, the whole strategy of physics when faced with a difficult problem often leads us to consider only the most essential aspects of the problem, even if the items eliminated from consideration are not necessarily trivial, but only of lesser importance. Thus we come down squarely on the side of the clarifiers and say, with some internal reservations, that living matter, as exemplified by a single cell, shows three important characteristics:

1. It is enveloped by a membrane structure and this structure is not a passive sack but a dynamic component of the system.

2. Chemical reactions occur from the membrane inward. These reactions provide energy, both for the various types of work done by the cell and for the production of some of the particular molecules required for the functioning of the system.

3. Information sufficient to permit the synthesis of required substances is stored as a molecular template in such a way that the living matter may reproduce itself.

Furthermore, we recognize that groups of cells that exhibit specialized features and properties exist in the form of tissues and organs; we assume that these features and properties are understandable in terms of the properties and features of single cells. Of course, it should be clear that these simple statements conceal very complex problems whose detailed solutions are great and as yet unrealized goals.

The identification and the determination of the precise composition of those molecules that are assembled to form living matter is one of the goals of biochemistry. From such studies, we have come to realize that the variety of molecules present in even a single cell is quite amazing. For example, a single cell of the bacterium $E.\ coli$ is a cylinder about 3 μm long, with a radius of about 1 μm and a mass of 10^{-12} to 10^{-13} g. As Lehninger has pointed out this cell contains about 5000 different organic compounds; about 1000 of these are different nucleic acids and about 3000 are different kinds of proteins. In comparison, a human being has about 5×10^6 different kinds of proteins and, as far as we know, none of the protein molecules in $E.\ coli$ is exactly the same as any one of the protein molecules found in human beings.

From this, it appears clear that the properties of the living state as a whole do not depend on specific individual molecules but rather on the organization of certain classes of molecules. How many different classes of molecules are needed? At first glance, this might appear to be a hopeless question. After all, as Figure 0.1 shows, the complicated structure of even a single cell is relatively obvious. There is certainly no reason to guess that this structure has its base in the properties of a relatively small class of organic compounds that are themselves composed of the members of a relatively small group of atoms. However, this turns out to be the case. Indeed, the existence of these "biochemical universals" greatly simplifies the physicist's task, because an enumeration of the immense number of different specific molecules becomes unnecessary; all that is required is a knowledge of the common features of the different categories. Let us now summarize the most elementary aspects of living matter, taking as an example a single unspecialized cell.

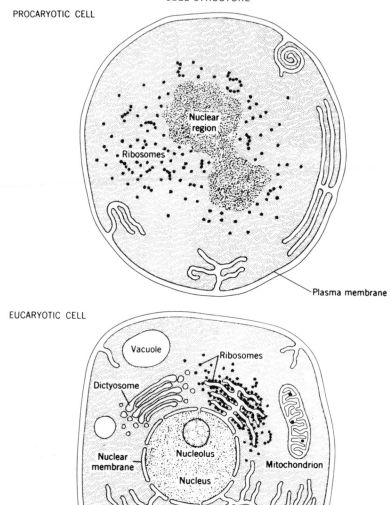

Figure 0.1 The major components of the two general types of cells are shown in the above drawings. The essential difference between the procaryotic and eucaryotic types is the presence of membrane-bound organelles, particularly the nucleus, in the eucaryotic cell. Bacteria and blue-green algae are procaryotic; the cells of all animals and higher plants are eucaryotic. Eucaryotic plant cells are distinguished from animal cells by a cellulose cell wall, chloroplasts for photosynthesis, and a central vacuole (see Figure 9.3).

Table 0.1 Critical Molecules

Compound	Comment	Example
Hexose sugars	Monosaccharides are compounds with the equation $(CH_2O)_n$, $n \geqslant 3$. Hexose sugars, $(CH_2O)_6$, are the most abundant monosaccharides.	Glucose, fructose
Trioses	The simplest monosaccharide, with the formula $(CH_2O)_3$.	Glyceraldehyde
Fatty acids	Long hydrocarbon chains terminated with COOH, the carboxylic group.	Palmitic acid, stearic acid
Purines	Complicated structures, formally known as nitrogenous bases.	Adenine, guanine
Pyrimidines		Thymine, uracil, cytosine
Steroids	Derivatives of compounds with three fused cyclohexane rings.	Cholesterol
Hydrocarbons	HC chains.	Squalene
Amino acids	Linkage of COOH, NH_2, and side chains.	Leucine, glycine

The key structural atoms from which living matter is built up are C, H, O, N, S, and P. These atoms occur more frequently in living matter than one might guess from their terrestrial abundances. The essential ancillary atoms are Na, K, Mg, Ca, Fe, Co, Mn, Cl, Cu, and Zn; these are less abundant in living matter than one might guess. Green has summarized the above situation very well: "There is no known case in which any of these atoms is replaced by an atom which is not part of the list. There may be additions, but never subtractions ... (these) particular atoms are invariant for all forms of life."

These atoms occur in living matter as ions (e.g., K^+, Na^+), as components of simple molecules (e.g., H_2O), but most especially as the components of certain critical types of molecules, listed in Table 0.1. These molecules are themselves the principal units from which the important macromolecules — the proteins, the nucleic acids, the polysaccharides, the lipids and phospholipids — are built up.

The construction of a protein molecule is based on linking amino acids, compounds formed by linking a carboxylic group, COOH, an amino group, NH_2, and certain specific radicals called side chains, as shown in Figure 0.2. The particular radicals that are linked determine the specific amino acid formed. All proteins are combinations of the 20 common amino acids. Although animals cannot synthesize all twenty from the simple molecules listed in Table 0.1, higher plants can, and hence animals can obtain these necessary amino acids.

The amino acids of a protein are linked to form a chain by means of a peptide bond, $\overline{-COCO-NH-}$, as shown in Figure 0.2. If the chain contains less than about a hundred links, it is usually called a peptide. If all the amino acids are the same, it is a

1. Amino acid structure:

$$R_1 \diagdown \atop R_2 \diagup C \diagup \mathllap{}^{COOH} \atop \mathllap{}_{NH_2}$$

which often appears as a polar
combination of two ions

$$^+H_3N-\underset{\underset{H}{|}}{\overset{\overset{R}{|}}{C}}-COOH^-$$

2. Amino acids are linked to form chains by a peptide bond:

$$R_1 \diagdown \quad R_2 \quad R_3 \diagdown \quad R_4$$
$$NH-\overset{|}{C}-CO-NH-\overset{|}{C}-CO$$

Peptide bond

3. Nucleic acids are formed by linking mononucleotides. Mononucleotides are formed from a combination of phosphoric acid, a nucleotide, and a pentose sugar. The combination of a pentose sugar and a purine or a pyrimidine is called a nucleoside. A nucleotide is formed by a PO_4 attached to the sugar of the nucleoside. The compound is then an acid, named by the nucleoside.

A. A Nucleotide

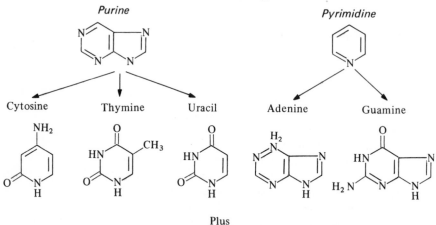

Purine   Pyrimidine

Cytosine   Thymine   Uracil   Adenine   Guanine

Plus

B. A Pentose Sugar

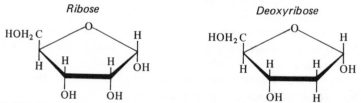

Ribose   Deoxyribose

Figure 0.2 This drawing summarizes the general structural basis of proteins and nucleic acids.

polypeptide. Proteins are usually considered to be one or more chains of from 100 to 300 various amino acids.

A second important class of macromolecules is the nucleic acids. These are formed by linking mononucleotides, which are formed as shown in Figure 0.2.

A third group, the polysaccharides are built up in a straight forward way from identical repeating units of glucose or some more complex sugar. The complex sugar that forms the repeating unit may itself be formed by the linkage of simpler sugars.

Finally, the last category, the lipids, are formed from the combination of fatty acids, and usually take the form of long chain molecules of fatty acid with glycerol. An important variation replaces one of the fatty acid chains by phosphoric acid, producing a phospholipid.

It is principally the above macromolecules that are organized to form the cellular components. What are the general roles of each category of macromolecule in the cell? The proteins play a variety of roles. They are enzymes, catalysts for reactions in the cell; they may serve as structural elements; and they are the important component in the mechanism by which transport of certain substances across the cell membrane occurs. The nucleic acids are essential for the storage of the information that permits the cell to reproduce itself, and they also play an important role in the energy processes of the cell. The lipids and phospholipids are the critical components from which the cell membranes are formed.

The general molecular components of a cell are summarized in Table 0.2. We can now show the structural organization of an "elementary" cell. Two general forms occur. In the first, the nucleic acid is not confined to a clearly distinguishable region of the cell, nor does it occur in combination with protein. Such cells, known as procaryotes, are typical of bacteria and blue-green algae. In all other cases, the nucleic acid is complexed with protein and confined to a well-defined region of the cell known as the nucleus. In addition, such cells, known as eucaryotes, contain a variety of specialized structures, also well defined, known as organelles, which apparently carry out specific functions. These two general cell categories are illustrated in Figure 0.1.

Of course, these figures do not exactly represent any particular type of cell, but only a useful generalization. Real cells are specialized in varying degrees. In some cases,

Table 0.2 Typical Bacterial Cell Composition

Component	Percent Total Weight	Number of Each Kind
H_2O	70	1
Protein	15	~3000
DNA	1	1
RNA	6	~1000
Carbohydrates	3	~50
Lipids	2	~40
Various molecules	2	~500
Inorganic ions	1	~12

such as the exocrine cells in the pancreas, the cells in the proximal tubule of the kidney, or the pallisade cells in a leaf, the specialization is not particularly extreme. In other cases, such as nerve cells or muscle fiber cells, considerable specialization has obviously occurred, to the point that many features of such cells would be unintelligable without more information than appears in the figures. Of course, there are still common features; for example, many of the organelles such as the mitochondria are similar over a variety of specific types of cell. It is clear that the cell occupies a central position both as a general structural unit, especially for simple life forms, and as the basis for specialization in both structure and function. It should be noted that cellular specialization is a stable phenomenon. Using techniques of tissue culture, pioneered by R. G. Harrison, cardiac cells, retinal pigment cells, and cartilage cells have been cloned for some 50 cell divisions without loss of specialization.

Within cells occur the wide variety of chemical reactions required for the production of energy, the synthesis of required compounds, and reproduction of the cell. It is probably true that nearly every organic compound can be used by some organism as an energy source. Clearly, there is no hope of simply summarizing such possibilities, and no attempt to do so will be made here. The immensely detailed processes of metabolism and biosynthesis have been revealed by decades of biochemical investigation and those interested in these processes should consult the standard texts.

Nevertheless, it is possible to summarize the elementary features of protein synthesis. The determination of the protein synthesis mechanism was a step of the greatest importance because proteins have fundamental roles in all of the essential cellular processes. The general problem is clear: the base sequence of the DNA specifies a set of amino acids. Therefore, there must be a process that translates this code in the form of base sequences into the set of amino acids that are linked to form the protein. In eucaryotic cells, tracer studies show that the proteins are produced in the cytoplasm. Since the DNA is in the nucleus, protein synthesis obviously does not occur on the DNA molecules. It seems clear that this is a general result for both types of cells. The process by which protein synthesis is accomplished is illustrated in Figure 0.3. This process is conveniently expressed in the so-called "central dogma": DNA → RNA → Protein. Although Temin and Baltimore have shown that certain viruses can interact with cells and carry out processes that do not follow this rule, the principle applies to all normally functioning cells.

The results shown in the figures serve to emphasize that DNA, RNA, and protein are key substances and cooperate in the production of other cell protein. The capacity for cellular protein production can be estimated by considering a bacterial cell. The DNA in such a cell consists of about 10^6 nucleotides. An average protein consists of approximately 100 amino acids. Since three nucleotides are required to specify an amino acid, 10^6 nucleotides code for some 330×10^3 amino acid molecules, which is about 3300 different protein chains.

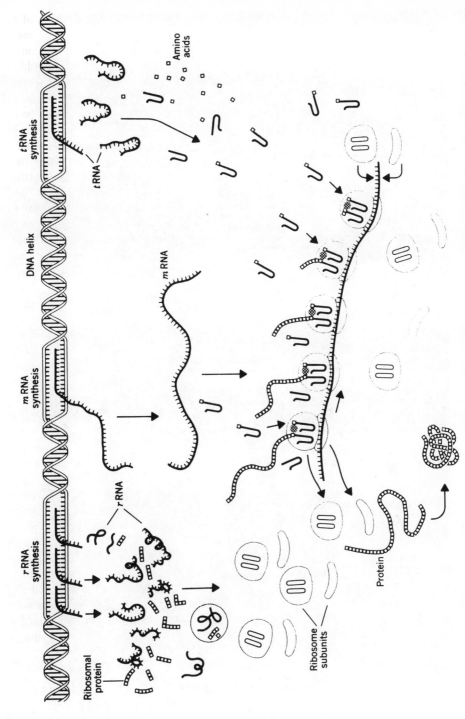

Figure 0.3 The general mechanism of protein synthesis appears to be essentially the same in all cells. The above figure shows, in a schematic way, the major steps in the process.

9

chap. 1

X-Ray Diffraction And Electron Microscopy

The biophysicist begins with a knowledge of the composition of biologically important molecules. An important goal is then the determination of the arrangement of the atoms in those particular molecules. Empirical rules from organic chemistry sometimes enable one to make correct guesses about the structure of various parts of a molecule. However, there is really only one way to determine directly and accurately the atomic arrangement for the molecule as a whole, and that is through X-ray diffraction, one of the classic techniques of physics. The importance of X-ray structure determinations in biophysics stems from a general conviction that if we know the functional role of a biologically significant molecule, then we can discover how the structure relates to that function. In addition, X-ray diffraction results also provide direct proof for many of the structural rules of organic chemistry.

THE MECHANISM OF X-RAY DIFFRACTION

The discovery of X rays occurred in 1895, when Roentgen noted that an "invisible radiation" that fogged photographic plates was produced by placing a small metal target inside a gas discharge tube. In 1912, von Laue, Fredrich, and Knipping realized that X rays were very short-wavelength electromagnetic radiation and could be diffracted by the closely spaced planes of atoms in simple crystals, in the same way that light was diffracted by the closely spaced lines of a diffraction grating. Bragg extended this work and showed that X-ray diffraction could be used to determine the arrangement of atoms in crystals; he also simplified the problem by demonstrating that the planes of atoms in such crystals acted like X-ray reflecting surfaces. He derived a simple and classic expression for the effect of considering a beam of X rays incident at some angle θ on a set of planes separated by a distance d. In Bragg's model, the successive similar planes of atoms in the crystal behave as an array of partially reflecting mirrors. The portion of the beam reflected from the first plane will be reinforced by reflection from the other parallel planes whose distance from the first, as measured along the direction of incidence, is an even multiple of the wavelength of the X ray. Hence the X-ray reflections are constructive, or reinforced, if

$$n\lambda = 2d \sin \theta \qquad (1.1)$$

as is demonstrated in Figure 1.1.

The validity of Bragg's argument can be shown experimentally in either of two ways. In one method, a simple crystal may be illuminated with a wide wavelength range of X rays. In this band of X rays, there will be one wavelength, λ, with precisely the value required to satisfy Equation 1.1, and a strong reflection will be detected. In the other method, the crystal is rotated while it is illuminated by X rays of a particular wavelength, which can be obtained by proper choice of the potential across the X-ray tube and the target material. Intense reflections will be found at the angles that satisfy

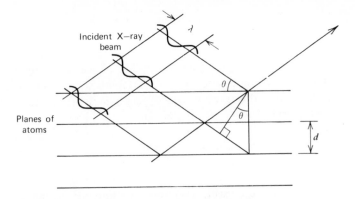

Bragg diffraction: atomic planes as partially reflecting mirrors

Figure 1.1 This diagram shows how the Bragg equation arises. The incident X-ray beam is partially reflected from a set of parallel atomic planes. The path difference is clearly $2d \sin \Theta$. Hence, the outgoing X rays will reinforce if the path difference is a multiple of the wavelength of the X ray. Thus, $n\lambda = 2d \sin \Theta$.

Equation 1.1. Of course, if the crystal is at all complicated, there will be many possible planes of atoms and thus many values of θ at which intense reflections will occur.

In simple terms, the reflection or scattering of the X rays is produced as follows. The X rays are short-wavelength electromagnetic waves. The oscillating electric vector of the electromagnetic wave produces an oscillating force on the electrons in the unit cell:

$$\mathbf{F} = e\mathbf{E} = e\mathbf{E}_0 \cos \omega t = m\ddot{x}$$

The electrons are, so to speak, shaken back and forth, and such accelerated charges radiate. If enough such processes are occurring, there will be radiation in a given direction whose electric vectors will be in phase and additive; an intensity maximum will then be produced. The intensity of the reflection will increase as the number of electrons participating in the process increases. Thus atoms like hydrogen are extremely poor scatterers. In fact, H is such a poor scatterer that the position of H-atoms cannot be found from X-ray methods.

A crystal must, of course, show some kind of three-dimensional pattern, and the deduction of this symmetry pattern from the X-ray diffraction is a critical step in the analysis of structure. Suppose we focus our attention on the repeating pattern of the crystal. Any set of similar points in this repeating pattern belongs to a lattice; picking other similar points simply shifts the lattice without changing its shape. The axes of this space lattice can be found by joining a point in the lattice to three neighboring points so that the set is not coplanar. The axes found in this way are the edges of the unit cell. The unit cell is then the arrangement of atoms that when regularly repeated in three dimensions builds up the extended, internal structure of the crystal. It is the dimensions of the unit cell that determine the directions (θ's) of the intensity maxima.

We can now add to the qualitative description of the scattering process given earlier: What X-ray diffraction procedures really establish is the electron distribution in the unit cell. Of course, this is really not a difficulty in itself; the atoms are located at the places where the electron density is greatest.

If we have many unit cells precisely arranged in the whole crystal, which is equivalent to saying the crystal is perfect or nearly so, then the X-ray reflections from the various unit cells will all be in phase. Thus the intensity of a given maximum increases with the crystal order. Furthermore, if the crystal is ordered, then the bigger it is, the more unit cells are present. Therefore, the intensity of the reflection also increases with crystal size, all other things being equal. In addition, the angular width of the maximum, or its spread about the mean value of θ, will decrease with crystal size; that is, the maximum becomes not only more intense but also more sharply defined with increasing crystal size.

If all the electrons associated with a given plane were exactly in that plane, then intensity variations as a function of angle would be the sole effect. However, the electrons are distributed about the nucleus and thus in or out of the plane. This leads to a slight change in the exact position of the maxima of the reflected electric-field vectors with respect to the maxima in the incoming waves; there is a slight phase shift in the reflected waves, due to the fact that, in effect, the reflection planes are not perfectly smooth. However, this phase shift, which can be an important source of information about the crystal, affects only the intensity of the reflection, not its direction.

ANALYSIS OF X-RAY DIFFRACTION PATTERNS

The goal of early X-ray crystal studies was the determination of the lattice size — the dimensions of the unit cell. Since many of the molecules of biological interest can also be crystallized, or at least teased into a semblance of order, it was natural to extend X-ray diffraction studies to these materials. However, in passing from relatively simple crystals to those of extremely complex molecules, we pass from relatively simple unit cells to unit cells that often contain thousands of atoms of various kinds. This leads to a change of emphasis. We are no longer interested so much in the dimensions of the unit cell as in the arrangement of the atoms within it. The goal of X-ray diffraction studies is now the explanation of the observed intensities in terms of the electron distribution in the unit cell. The mathematical basis for this procedure may now be summarized.

We begin by noting a crystallographic convention for locating planes of atoms in crystals. It has been shown that there is always a set of axes such that when the reciprocals of the intercept coordinates are taken, these reciprocals can be expressed as a set of integers. These integers are known as Miller indices; they were first used in crystallography in the 1840s and are universally designated by the letters h, k, and l. The physical positions of planes defined by different sets of indices is shown for several examples in Figure 1.2, which should make the physical meaning of the indices clear.

Let us now find the connection between the crystal structure and the intensities of the various spots in the X-ray diffraction pattern. The incident X ray may be described by the equation:

$$\mathbf{E} = \mathbf{E}_0 \exp \beta \qquad \text{where} \qquad \beta = i \left(\frac{2\pi}{\lambda} x - \omega t \right)$$

and where $\mathbf{E}$ is the instantaneous field, and $\mathbf{E}_0$ the maximum electric field strength; λ is the wavelength, and ω is the angular frequency. This E exerts a force on each

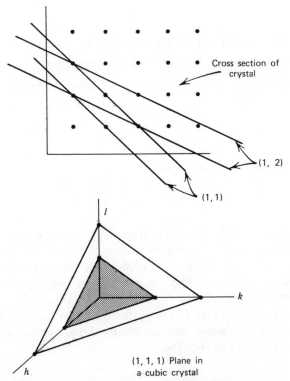

Figure 1.2 The above figure indicates how specific planes of atoms in a crystal may be indicated by specifying the Miller indices.

electron in the unit cell, given by

$$\mathbf{F} = \mathbf{E}e$$

which produces an acceleration:

$$\mathbf{a} = \mathbf{F}/m = \mathbf{E}e/m = e/m\ \mathbf{E}_0\ \exp\beta$$

These accelerated charges radiate. An essential point is that although the direction of the radiation produced is independent of any phase difference between the incident and the diffracted beam, the intensity will depend on the phase difference.

Suppose that f_s is the amplitude of the wave scattered by an atom and ϕ_s is the phase. Then we define the structure factor:

$$F = \sum_n f_s e^{i\phi_s}$$

where n is the number of atoms in the unit cell. It can be shown that

$$e^{i\phi_s} = \exp\left[2\pi i\left(\frac{h}{a}x_s + \frac{k}{b}y_s + \frac{l}{c}z_s\right)\right] = e^{2\pi i f(\text{coordinates})}$$

where

> h, k, and l are Miller indices
> a, b, and c are the axial lengths of the unit cell
> x_s, y_s, and z_s are the coordinates of the scattering atom

In general, f_s depends on the particular atom (e.g., the more electrons, the more scattering) and the angle between the incident radiation and the scattering plane. From biochemical results, the composition of the molecule is known and the angle is determined in the experiment. This, in combination with previously available laboratory data, enables us to calculate f_s values.

Since it is the electrons that do the scattering, a realistic model of the crystal is one with a continuous but variable electron density throughout the unit cell. Let the electron density distribution be $\rho(r)$. Then

$$F = \int_{\substack{\text{unit} \\ \text{cell}}} \rho(r)e^{2\pi i f(\text{coordinates})} \tag{1.2}$$

Physically, $|F|$ is the ratio of the amplitude of the wave diffracted by the unit cell to the amplitude diffracted by a single electron. Equation 1.2 may be inverted to give $\rho(r)$ in terms of a Fourier series:

$$\rho(r) = \Sigma |F| \cos[2\pi f(\text{coordinates}) + \alpha] \qquad (1.3)$$

were α is the phase shift produced by the diffraction; thus, in order to get $\rho(r)$, we must find both F and α.

The intensity of the scattered radiation is proportional to $|F|^2$. Determining the phase angle α depends on being able to diffuse some complex containing a heavy metal atom such as I, Br, or Hg into the crystal in a more or less known position. Heavy metal atoms have, relative to the atoms common in biological molecules, many electrons. Therefore, they are very efficient X-ray scatterers and diffract much more strongly than the other atoms in the molecule. Hence, to a good approximation, $f_s \doteq 0$ for all the other atoms, and this means the only unknown is α. Once we have a single set of approximate values, we may give up the approximation and find the true values of F and α by iteration. This technique was invented by Perutz in his work on hemoglobin (MW 64,500); the first molecule "solved" using this method was myoglobin (MW 17,500).

The labor of this task is clear. There may be thousands of atoms in the unit cell and Equation 1.3 will contain one term in the sum for each set of planes hkl, that is, one term for each reflection spot in a perfect X-ray diffraction picture. One may choose to measure fewer spots than are present, but then the electron density function is less well determined and resolution is lost; resolution is increased by measuring reflections from larger angles. For example, in the case of lysozyme, the analysis of which is illustrated in Figure 1.3, a low-resolution, electron density map can be obtained from the measurement of 800 reflection spots. However, in order to "see" amino acid side chains as distinct structures, a resolution of .2 nm is required, and this means that about 20,000 reflections must be measured. In the case of myoglobin, 400 reflection measurements are required for a resolution of .6 nm; 10,000 for .2 nm; and 25,000 for .14 nm. Thus, to double the resolution, the number of measured reflections must be increased by a factor of eight. Furthermore, in the case of the .2-nm resolution, we need 10^4 reflections for the native protein, plus 10^4 for each case of heavy atom substitution. For each set of Miller indices, we require 10,000 equations from which the F-value will yield the phase angle. The phase and amplitude values then define the 10,000-term Fourier series, which yields the values of the electron density distribution in the unit cell. To get a .14-nm resolution, the Fourier sum was computed for 500,000 points in the unit cell.

Does the need for high resolution imply that a large crystal is needed? No, a quite small crystal will still contain many unit cells. We can calculate an appropriate minimum size in the following way. If the angle satisfies Equation 1.1, the amplitudes of the reflected waves add; this can be represented vectorially (Figure 1.4). We want to

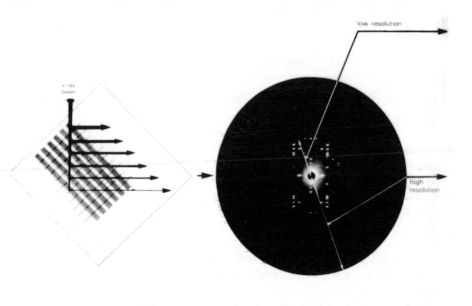

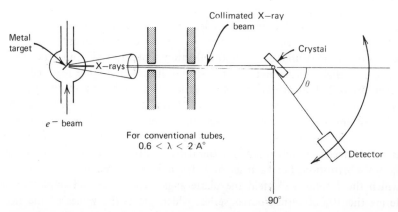

Figure 1.3 The general procedures for X-ray structure determination are illustrated above. A molecular model is the end product of the effort. In this case, it is a model of lysozyme, but the above procedure is of course the same for other molecules. The general arrangement of the X-ray spectrometer is shown at left, and diffraction by the crystal is indicated schematically. A typical diffraction pattern is shown at the lower right. The final result of the analysis depends on whether the data are adequate for a high- or low-resolution model; the distinction between these results is clear from the two models shown at the upper right.

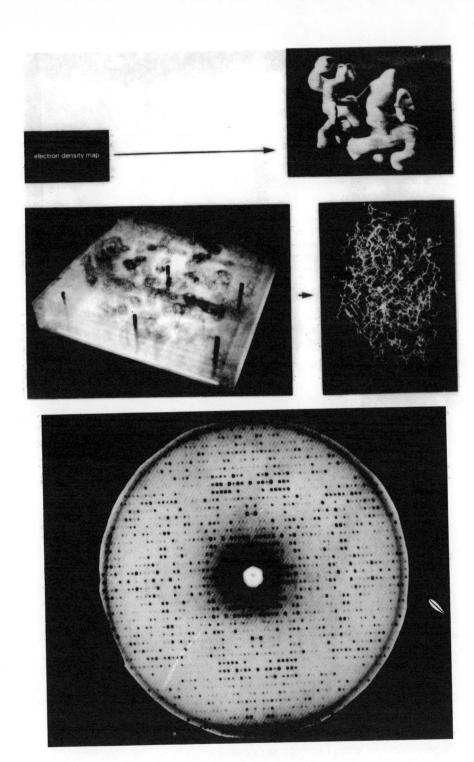

electron density map

19

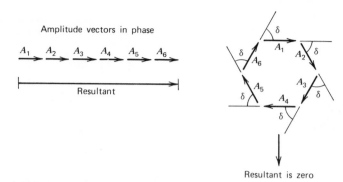

Amplitude vectors in phase

Resultant

Resultant is zero

Figure 1.4 The above shows the two extremes for the addition of the amplitudes. The resultant intensity goes to zero if the path difference is λ/N; thus the amplitude vectors form a closed figure. Note that the figures show a small number of vectors purely for convenience; actually, there is one vector for every plane.

find $\Delta\theta$ as a function of the number of reflecting planes, N. Suppose there is a slight variation in θ. Then the reflected waves are out of phase by some small angle δ. If the path difference, or separation of the reflecting planes, is λ/N, the waves will be exactly out of phase and the resultant will be zero. Thus the vector diagram will close, as shown in Figure 1.4. Therefore,

$$\delta = 2\pi/\lambda \; \lambda/N$$

so

$$n\lambda + \lambda/N = 2d \sin(\theta + \Delta\theta)$$

whence

$$\Delta\theta = \lambda/2Nd \cos\theta$$

and the total angular width of a reflection will be

$$2\lambda/Nd \cos\theta = 2\lambda/t \cos\theta$$

since the number of planes, N, multiplied by their separation, d, is just the thickness of the crystal, t.

If we now substitute the usual minimum obtainable line width for an X-ray diffractometer, we find that it equals the total angular width of a reflection when the thickness of the crystal t is about 10^{-4} cm. Of course, this crystal, though obviously small, is still quite large on a molecular scale.

The earliest important X-ray results for biological molecules came from Astbury's studies of proteins. He showed that many protein X-ray patterns could be placed in one of two categories. Either they were tightly folded up, in the so-called α-configuration, or stretched out, in the β-configuration. Astbury was also able to show that in some cases stretching material with a nominally α-pattern produced a β-pattern. Somewhat later, Corey and Pauling showed, on chemical grounds, that proteins could be twisted into a helix, the α-helix, with 3.7 amino acids per turn. This gave a helix with a spacing between the turns of .54 nm and a diameter of about .6 nm. Cochran, Crick, and Vand were then able to show that this model gave an accurate prediction of observed X-ray patterns at low resolution.

This work laid the foundation for the first studies of proteins at a resolution sufficient to reveal the detailed arrangement of the amino acid chains. The first high-resolution results were rather surprising; instead of regular structure, one found, in Kendrew's words, "almost nothing but a complicated set of polypeptide rods, sometimes going straight for a distance, then turning a corner and going off in a new direction ... much more complicated and irregular than most of the early theories of protein structure had suggested."

These complicated protein structures, some of which are shown in Figure 1.5, have been amply confirmed by later studies at even greater resolution. Although structural similarities between proteins with different sequences but similar functions are now known, it is reasonable to ask if there are any regularities in overall protein structure. The answer that is now emerging appears to be a qualified "yes," and we will return to this matter in a later discussion of the physics of macromolecules. The major effort in X-ray diffraction analysis has been concentrated on the protein structure problem, both because of the importance of proteins and, so far, the absence of simple rules for determining the detailed structure. However, X-ray structure determinations have also been applied to other macromolecules of biological importance; the most celebrated example is the determination of the DNA double helix structure (Chapter 5).

Although the development of computers has greatly reduced the labor of the iterations needed to determine $\rho(r)$, X-ray diffraction studies of macromolecules are still far from simple tasks. These extensive computations lead to difficulties for the experimentalist who finds that a long time is required to arrive at the point of data production and then suddenly there is a flood of data. In practice, one copes by converting the numerical data into contour lines of equal electron density that are then drawn on clear plastic sheets. However, the model of the molecule is still to be produced. Originally, one went to a "Richards box," which is an optical comparator so designed that one could simultaneously view the tentative model of the molecule, usually made from wire, and the plastic sheets with the electron density contours. Adjustments in the model could then be made, seen against the known electron density contours. Improvements in this approach began with automated techniques, pioneered by Diamond, which allowed one to construct models based on coordinates of atoms deduced from the contour maps. This method was improved so that the

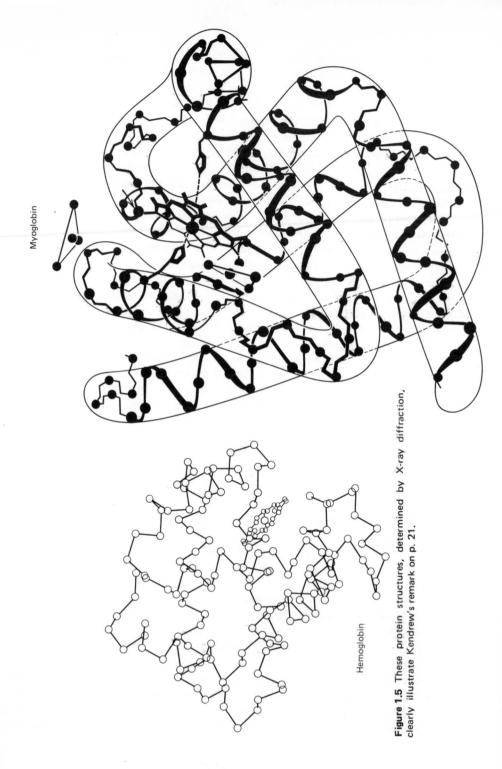

Myoglobin

Hemoglobin

Figure 1.5 These protein structures, determined by X-ray diffraction, clearly illustrate Kendrew's remark on p. 21.

model could be directly fitted to the contour map, and an iteration technique was developed that in effect used the model with the best fit to improve the values of the phase angle α required in Equation 1.3. This iteration technique was further improved by Diamond and Levitt, who added energy requirements for the various bonds and then required the model to fit these as well as the electron density map. In principle, an almost totally automated system is possible, with the X-ray diffraction apparatus "on line" to the computer, which controls the apparatus, records the data, analyzes it, and produces the result; recent work by Greer has shown how a complete analysis of the $\rho(r)$ map could be carried out without any "tinkering." Such developments are necessary if we want detailed structural information on even a fraction of the natural proteins, and such information may be medically important, since genetic defects can manifest themselves in small alterations in the order of the amino acids in the protein.

The methods of X-ray diffraction give us high-resolution "pictures" of biological molecules. Nevertheless, it is obvious that "real" biological structure is produced by assemblies of such molecules. Thus we also need to be able to examine structure on a scale appropriate to these assemblies and this is from a few tens up to a few thousand Angstroms. Although low-angle, X-ray diffraction can give some information about structures with characteristic dimensions in this range, the dominant technique is electron microscopy.

PHYSICAL BASIS OF ELECTRON MICROSCOPY

As can be seen from Figure 1.6, the electron microscope shares many parallels with the optical microscope, whose essentials are summarized in Appendix 2. The performance of either type of microscope can be described in terms of resolution and magnification. Of these, resolution is the more important because it determines the smallest detail that will be visible. Resolution varies inversely with the wavelength of the radiation used to examine the specimen; for the optical microscope, the limit of resolution is about 300 nm. Although this is sufficient to reveal the larger aspects of cellular organization, it is not even remotely equal to the task of assertaining the structural complexies of the cell.

In the early 1930s, the realization that it should be possible to construct a microscope with vastly improved resolution generated the impetus for the development of the electron microscope. The development of the electron microscope is a very good example of the difficulties of predicting the practical consequences of pure research; that development was in fact initiated by answers to fundamental questions about the nature of light and matter. It came about in the following way.

In 1912, one of the central problems of physics was the explanation of the photoelectric effect, which is the emission of electrons when light is shined on certain metal surfaces. The essential facts in the effect are as follows:

1. There is a critical wavelength, above which no electron emission occurs.

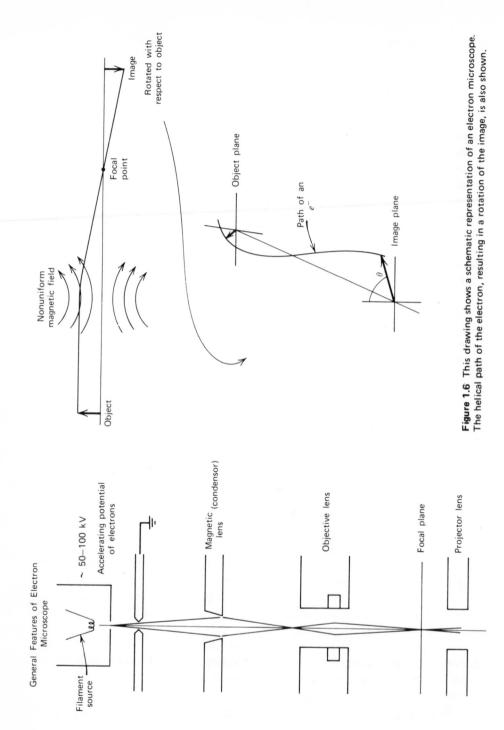

General Features of Electron Microscope

~ 50–100 kV
Accelerating potential of electrons

Filament source

Magnetic (condensor) lens

Objective lens

Focal plane

Projector lens

Nonuniform magnetic field

Object

Focal point

Image

Rotated with respect to object

Object plane

Path of an e^-

Image plane

θ

Figure 1.6 This drawing shows a schematic representation of an electron microscope. The helical path of the electron, resulting in a rotation of the image, is also shown.

2. The number of electrons emitted depends only on the intensity of the light.

3. The energy of the emitted electrons depends only on the frequency of the light.

Einstein explained all of these results, and others, by showing that light could be considered to behave as a beam of particles, now called photons. The energy of a photon was expressed in terms of the frequency used in the conventional description of light as an electromagnetic wave:

$$E = h\nu$$

This result marks the introduction of duality into physics: the realization that wave and particle descriptions are not mutually exclusive.

Louis deBroglie was profoundly influenced by this result; in 1925, he argued that if waves had particle aspects, then the symmetry of nature required that particles should exhibit wavelike properties under the proper experimental conditions. He deduced that a particle with a momentum p should also exhibit properties appropriate to a wave of wavelength λ through the following argument:

$$E = mc^2 \qquad \text{and} \qquad E = h\nu$$

so

$$h\nu = mc \times c$$

But, since any mass x velocity $\equiv$ momentum,

$$h\nu = pc$$
$$\therefore p = h\nu/c$$

But, since $c = \lambda\nu$,

$$p = h/\lambda$$

Experiments by Davison and Germer showed that electrons could be diffracted by crystal lattices. Because diffraction is a typical wave phenomena, deBroglie's view was confirmed.

Since the electron clearly possessed wave properties, it seemed reasonable to assume that the laws of optics applied. From

$$\lambda = h/P$$

it is clear that even at modest momenta, the wavelength of the electron is very much less than the wavelength of light. Thus, if a microscope could be made using electrons instead of light waves, the improvement in resolution would be very impressive. For example, a resolution of .002 nm would require a wavelength of .005 nm, and this could be produced by an accelerating potential of only 50,000 V, easily produced in the laboratory. Since the electrons do attain high velocities with these potentials, the momentum should be calculated from the relativistic expression:

$$p = m \cdot v/\sqrt{1 - v^2/c^2}$$

The kinetic energy of such an electron will be given by

$$T = m_0 c^2/\sqrt{1 - v^2/c^2} - m_0 c^2$$

The results of this quick calculation should not raise exaggerated hopes; although the wavelength of the electron is very short indeed, a variety of difficulties, some of which are insurmountable, prevent us from attaining more than a fraction of the resolution possible with such short wavelengths. Even so, that gain represents an immense improvement over the light microscope.

The problem of building a lens to focus the electron wave seemed straightforward. One possible approach is simply to subject the electrons to properly oriented electric fields. However, this approach has a number of difficulties, and most electron lenses employ magnetic focusing.

It is well known that a magnetic field exerts a force on a moving charge given by

$$F = q \, v \, B \sin \theta$$

where θ is the angle between v and B. It is equally well known that magnetic fields are produced by currents flowing in wire coils.

Now consider a long coil (i.e., a solenoid); B is uniform inside the coil. The electron will move down the axis of the solenoid in a helical path. The distance the electron travels in making one complete turn about the helix is easily shown to be

$$\frac{2\pi m}{eB} \, v \cos \theta$$

In order to produce a lens from this arrangement, we must now make the field nonuniform. This is the central problem of magnetic lens design: how do we wind the solenoid in such a way as to produce a nonuniform B that will magnify the image. We need to subject the electron to a force that increases with increasing distance from the axis of the solenoid. This can be done without too much difficulty by placing a properly shaped sleeve around the coil. However, the position of the image plane of the electrons will be proportional to $\cos \theta$, (i.e., the lens is subject to spherical aberration). Now

$$\cos \theta \simeq 1 - \tfrac{1}{2}\theta^2$$

So if we keep θ small, the distance the electron travels becomes independent of θ, and this minimizes the aberration. However, this means we need a very well-collimated beam. Since this is usually done by a series of apertures placed along the electron beam, it is obvious that the aperatures must be very small. That being the case, not many electrons, relative to the number produced at the filament, can get through, and therefore we have low intensity. Of course, we are also assuming that all the electrons can be brought to the lens system traveling at the same velocity; if this isn't so, then they have slightly different λ's. The lens system is not achromatic; it only works for one specific electron wavelength. Other wavelengths will, therefore, come to focus at slightly different positions; this is the electron lens equivalent of chromatic aberration.

There are other problems. One sees image detail in conventional microscopy because different parts of the specimen absorb different amounts of light, that is, image contrast is due to intensity variation. In the case of electron microscopy, the penetrating power of the electrons is very slight. Therefore, the specimens must be very thin. However, since the difference in electron absorption between various parts of the specimen is minor there is no contrast between the different parts of the specimen and the whole image is a vague, essentially featureless region.

This problem has an analogue, although it is less serious, in optical microscopy. In that case, to solve the problem, we stain the specimen using dyes that we know, by trial and error, have a different affinity for different cell components. This greatly enhances the contrast between different parts of the cell. A similar solution has been adopted in electron microscopy, but the stains are now solutions of heavy metals like osmium tetroxide or uranyl acetate. The metal is taken up differently by different cell components. Thus different regions of the cell are now more or less effective scatterers of electrons, depending on whether they have taken up more or less of the heavy metal stain. In this way, contrast between the different cell components is produced artificially. The preparation of specimens for examination by electron microscopy is illustrated in Figure 1.7.

Of course all such staining is done on dead material, and we can wonder what changes in structure may occur because the material has been killed, and what changes may be produced because it has been subjected to these none too gentle chemical preparations.

Another approach to the problem of low contrast in electron microscopy is to

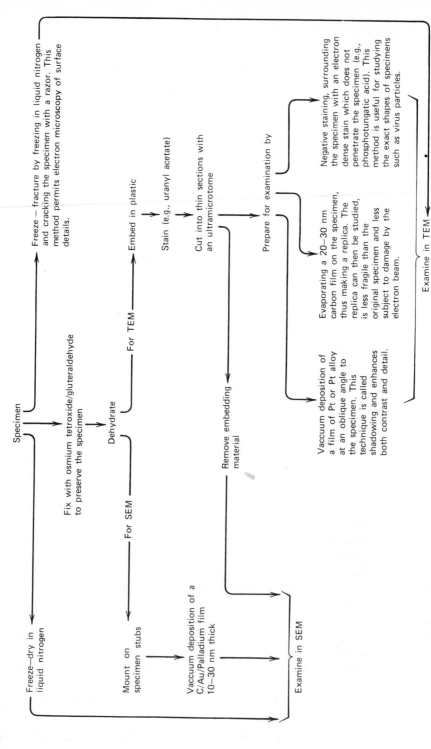

Figure 1.7 The above figure shows, schematically, the preparation of specimens for electron microscopy.

28

employ various techniques for intensity amplification and image processing. The requirement for thin sectioning, however, can only be circumvented by using a higher accelerating potential. If this were possible, the use of thick specimens, perhaps in some cases living, would be possible, and the increase in specimen thickness would lead to more effective energy dissapation and hence less specimen damage by the beam. If the potential could be raised to ~750 to 1000 kV, then essentially any specimen of reasonable thickness could be examined. Of course, the contrast problem would be worse than at present and, consequently, intensity amplification and image processing would be that much more important. The limiting factor in the case of such a high-voltage microscope is the inability to improve the performance of the electron lenses. Although the specimen would transmit ~10% of the incident beam, only ~1% could be collected by the lens, and the effective f-number of the system would be ~500. The principal design problem at this time is how to reduce the spherical aberration of the small-aperature, electron lens.

It is the spherical aberration of the lenses that sets the practical resolution of the electron microscope. If we are close to the optical axis, spherical aberration and diffraction both contribute to fixing the resolution. Furthermore, the contribution of the projection lens is far less significant than that of the objective. If we choose a value of θ so that the errors due to diffraction and aberration are about equal, then

$$\theta \approx \left(\frac{0.6\lambda}{K}\right)^{\frac{1}{4}}$$

where K is a lens constant that is proportional to the spherical aberration. The minimum resolvable separation can then be shown to be

$$d_0 \doteq 0.7\, K^{\frac{1}{4}} \lambda^{\frac{3}{4}}$$

This is consistent with Abbé's principle: A grating with spacing d_0 cannot be resolved unless the lens aperture accepts both the $n = 0$ and $n = 1$ maxima. For an electron microscope objective lens with $f = 0.15$ cm, K is ≈ 0.05, so $d_0 \approx .5$ nm. This is, of course, far less resolution than one would expect from arguments based solely on the deBroglie wavelength of the electron.

The above discussion assumes that nothing is happening to the specimen under observation; thus what is derived is the practical, physical limit to the resolution. There are also biological limits to the resolution due to the heating of the specimen in the electron beam when being studied. Other effects, such as the diffusion of heavy metal stain and the distortion of the structure during specimen preparation, occur but it is hard to be quantitative about these. In any event, it is clear that such effects make the study of structures smaller than about 2 nm very risky and many would suggest a limit closer to 4 nm.

chap. 2

Some Other Methods For Studying The Components Of Living Matter

The complete structural characterization of biological macromolecules by X-ray diffraction is not soon likely to become an everyday laboratory technique. Fortunately, in many cases, a less detailed knowledge of these molecules, as well as of certain other cell components, if often sufficient. In such cases, there are a variety of methods for obtaining particular pieces of information. These methods may be summarized as follows:

1. Centrifugation can be used to get the molecular weights of macromolecules, as well as to separate both macromolecular components of a cell (e.g., the proteins from the nucleic acids) and cellular components (e.g., the membrane structures from the nuclei).

2. Spectroscopy can be used to reveal the changes that may occur in molecular configurations in response to various conditions. In addition, the size and, in some cases, the actual shape, of macromolecules may be determined. Spectroscopy includes a variety of techniques involving both conventional optical methods such as circular dichroism, optical rotation, and light scattering, as well as others, such as nuclear magnetic and electron spin resonance.

3. Electrophoretic techniques can be used to separate charged macromolecules, as well as to measure their charge.

4. Viscosity measurements can provide information on macromolecular sizes.

CENTRIFUGATION

In physics, one of the fundamental pieces of information about an object is its mass; macromolecules are not exception to this rule. As Newton's second law:

$$F = \dot{P} \qquad \text{where} \qquad P = mv$$

shows, masses are determined by observing their motion when acted on by known forces.

$$m = F/a \qquad \text{if} \qquad dm/dt = 0$$

The most obvious force is gravity, but unfortunately this force is too weak to use in a direct determination of the mass of a macromolecule. This can be seen from the following calculation. Consider a light macromolecule with a molecular weight of

10^4 Daltons; its mass is given by

$$m = \frac{\text{molecular weight}}{\text{Avogodro's number}} = 1.6 \times 10^{-23} \text{ kg}$$

Suppose we have two such macromolecules in a test tube; let us say that they have a vertical separation of 1 cm. What is the difference in their potential energies? Let the potential energy be denoted by U. Then by definition

$$U = - \int_0^{1 \text{ cm}} dW \qquad \text{where} \qquad dW = Fdx$$

The force in this case is gravity and, therefore,

$$F = m \frac{M_\oplus G}{R_\oplus{}^2} = mg$$

$$\therefore U = - \int_0^1 F dx = -mg \int_0^1 dx = mgh = 1.6 \times 10^{-23} \text{ kg} \times 10 \text{ m/sec}^2$$
$$\times 10^{-2} \text{ m}$$
$$= 1.6 \times 10^{-24} \text{ J}$$

How does this compare with the average thermal energy of this macromolecule? By definition,

$$\langle \tfrac{1}{2} mv^2 \rangle = \tfrac{3}{2} kT = \bar{E}_{\text{thermal}}$$

so

$$\bar{E}_T = \tfrac{3}{2} \times 1.38 \times 10^{-23} \text{ J} \times K \times 293° \text{ K}$$
$$= 4 \times 10^{-21} \text{ J}$$

Therefore, since

$$\frac{\text{Thermal energy} =}{\text{Potential energy} =} \ \cdot \ 2500$$

the thermal energy of the macromolecule at $0°$C is about 2500 times larger than the gravitational contribution. Thus the motion of the particle is completely dominated by its thermal motions, which are totally random. What is the solution to this difficulty?

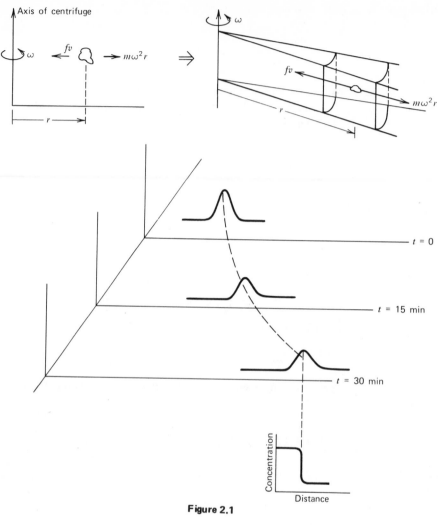

Figure 2.1

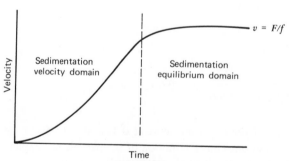

Figure 2.2

34

Obviously, we need to increase the "effect of gravity" until it is greater than kT. This was accomplished by Svedberg in the 1920s, who developed the high-speed centrifuge, in which a rapidly rotating container is used to exert a very large, artificial gravity on its contents.

The centrifugation process, illustrated in Figures 2.1 and 2.2, can be analyzed in the following way: The force per unit mass on a particle at x due to the rotation of the rotor at an angular velocity ω is

$$F = \omega^2 x$$

The centrifugal force on the particle is obtained by viewing the particle from a reference frame rotating with the system. The net force on the particle is the difference between the centrifugal force (as seen from the rotating axis) and the sum of the frictional force due to the particle's motion through the fluid and the buoyant force produced by the weight of the fluid displaced by the particle; that is,

$$m\omega^2 x - \frac{m}{\rho}\rho_0 \omega^2 x - fv = \text{net force on particle}$$

where ρ and ρ_0 are the densities of the particle and the suspending fluid, respectively, f is the frictional coefficient, v the velocity of the particle in the medium, and m is the mass of the particle.

Of course, this is not a very convenient expression. However, two experimental techniques can now be used; these are illustrated in Figures 2.3 and 2.4. The first of

A. Prepare a density gradient

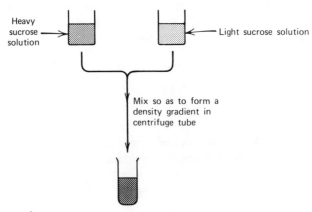

Heavy sucrose solution

Light sucrose solution

Mix so as to form a density gradient in centrifuge tube

B. Spin until $\dot{v} \rightarrow 0$ (long t)

C. Extract layers

Figure 2.3 Sedimentation equilibrium method.

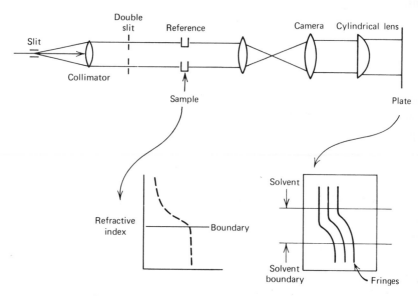

A. Measure average sedimentation velocity optically; e.g. with Rayhigh interferometer

B. Calculate sedimentation constant

Figure 2.4 Sedimentation velocity method.

these, the sedimentation equilibrium method, is very slow because the centrifuge rotor must be spun until the particle velocity in the medium goes to zero. The advantage of this method is that the particle masses can be determined independent of size or shape. In the second technique, optical methods can be used to measure the average velocity down the tubes. Since the only requirement now is to wait until the velocity is constant, this method is much faster than the first. In this sedimentation velocity method,

$$\bar{v} = \frac{m\omega^2 x}{f}\left(-1 - \frac{\rho_0}{\rho}\right)$$

The measurements give $\bar{v}$; the value of $\omega^2 x$ is determined by the choice of rotor and speed. We now define the sedimentation constant:

$$s \equiv \frac{dx/dt}{\omega^2 x} \equiv \frac{\bar{v}}{a}$$

where a value of 10^{-13} sec is said to be one Svedberg or S-unit. Sedimentation constants are usually defined as rates in water at $20°C$. Since the density of most of the materials of interest is essentially independent of T, the standard sedimentation constant, $S_{20,\omega}$, is given by

$$S_{20,\omega} = \frac{dx/dt}{\omega^2 x} \cdot \frac{\eta_{T,M}(\rho - \rho_{20,w})}{\eta_{20,w}(\rho - \rho_{T,w})}$$

where ρ is the density of the particle and $\eta_{T,M}$ and $\eta_{20,w}$ are the viscosities of the medium at the temperature of the centrifuge tube and the viscosity of water at $20°C$, respectively, with a similar notation for the densities $\rho_{T,M}$ and $\rho_{20,w}$. Now

$$m = s \frac{f}{1 - \rho_0/\rho}$$

The value of f can be determined from diffusion experiments, discussed in Chapter 6. Here we only note that the diffusion constant D can be measured and that

$$D = \frac{kT}{f}$$

Thus

$$m = s \frac{kT}{D(1 - \rho_0/\rho)}$$

which becomes an expression for the molecular weight if both sides are multiplied by Avogodro's number, N_A. The density of the particle, ρ, may be obtained by centrifugation in media of different ρ's; ρ is that ρ in which a layer of the material in question is not displaced. As the equation shows, the mass of a particle may also be expressed in S-units. If the particle is a sphere of radius r and the medium has a viscosity η, then the value of s may be calculated from

$$s = \frac{2r^2(\rho - \rho_0)}{9\eta}$$

An example of a very important question settled by centrifugation is provided by the Meselson-Stahl experiment, done in 1958. Crick and Watson had proposed the double helix structure for DNA in 1951. The question was how the DNA could replicate itself, and several possibilities were suggested. In order to settle the question, Meselson and Stahl labelled double-strand DNA with N^{15}, thus making it somewhat heavier than normal DNA, which contains only N^{14}. This was accomplished by growing *E. Coli* on a medium containing ammonium chloride made from heavy nitrogen. The bacteria took up the NH^{15}_4Cl, and some of the NH^{15} ended by being incorporated into the bacterial DNA. The bacteria were then shifted to a medium with normal NH_4Cl and the nucleic acid replication with time was investigated. DNA extracted from samples at different times was centrifuged in a solution of cesium chloride in which there was a density gradient, with the density ranging from 1.66 to 1.76 g/cm^3. The density of normal (i.e., unlabelled) DNA is about 1.7 g/cm^3, and the

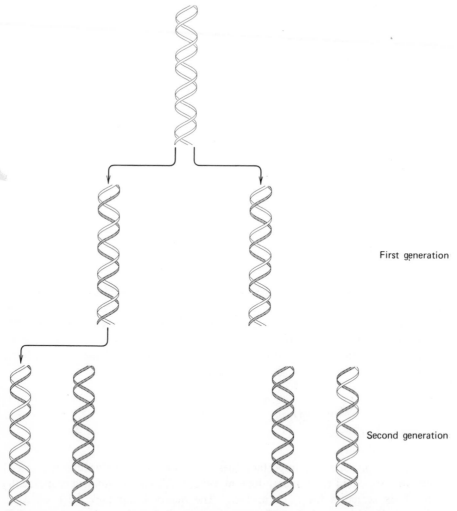

First generation

Second generation

Figure 2.5 The above sketch indicates the central idea of the Meselsohn–Stahl experiment, in which centrifugation of labelled DNA was used to determine the mode of DNA replication.

difference in density between the labelled and unlabelled forms was about 1%. If the bacterial DNA was taken after one generation time, that is, after it had replicated once inside the bacterial cells, centrifugation revealed that only one band of molecules was found in the centrifuge tube. The density of this band was halfway between that of the labelled and unlabelled DNA. In other words, the replicated double-strand DNA obviously had one heavy or labelled strand and one light or unlabelled strand. After two generation times, two bands were found. One was double-strand DNA with one N^{15}-containing strand and one N^{14}-containing strand. The other was composed of pure N^{14}-containing strands. Clearly, the steps in DNA replication were as shown in Figure 2.5.

Interest in centrifugation has developed steadily. Early centrifugal techniques focused on breaking up the cell into components. Once the problems of isolating subcellular organelles had been solved, the focus of the efforts shifted to producing specific organelle fragments and then to separating the various molecular components that compose these fragments. Roughly speaking each of these developments requires an increase in rotor speed; rotors with speeds of $\sim 6 \times 10^3$ r/min are required to separate particles in the size range $\sim 1 \ \mu m$, while the rotor speed must increase to about 50×10^3 r/min in order to separate particles in the size range $\sim 0.05 \ \mu m$. The present emphasis of reasearch in centrifugation, such as that of Anderson's group at Oak Ridge, is the development of very high-yield systems.

SPECTROSCOPY

In biophysics, the spectra produced by the interaction of light and matter provide information about the detailed structure of individual molecules, as well as clues that can be used to determine the physical state of the molecular components in various cell structures. Many of these applications of spectroscopy can be understood in a qualitative, or simple quantitative way, by choosing to emphasize, depending on the particular case, either the photon aspect or the wave aspect of the interaction. Although this choice of emphasis does appear quite arbitrary, it is really not, and is only a consequence of our desire to provide simple physical pictures. A consistent explanation of the production of spectra, without the need to make arbitrary decisions, is impossible. However, such an approach must be from the viewpoint of quantum mechanics, with its attendant mathematical complexities.

For our purposes, the essential quantum mechanical result is the demonstration that there is a restriction on the possible energies of an atom or molecule. The first indication of this result was obtained by Bohr for the case of the hydrogen atom. The problem was essentially solved by Schrodinger and Heisenberg, and a complete solution was obtained by Dirac.

The physical model for understanding the production of spectra is relatively simple. The system, be it atomic or molecular, has a set of permitted energy levels. A photon whose energy is equal to the difference between the energies of two particular levels in the system may be absorbed by the system. This places the system in a higher energy state. Conversely, a system that is in a particular energy state, other than the lowest possible one, may decrease its energy by emitting a photon. The energy of this photon will be the difference between the energy of the particular upper state and the lower state of the system after the emission.

Most of the spectroscopic problems in biophysics involve the study of absorption lines. This is because, in order to observe emission lines, we must first begin with the molecules of the sample (e.g., some protein molecules) in excited states from which photons can be emitted as the molecules decrease their energy. However, to get these molecules into excited states, we must first supply the necessary energy. Usually this is done by heating the sample; such heating destroys protein structure as well as other biologically important material with which it may be associated.

Absorption spectra are observed by illuminating the sample with a wide range of photon energies. Those photons whose energies are equal to the energy difference between two states may be absorbed while the rest will not be absorbed. Thus if we look at our beam of photons after the interaction with the sample, those that were absorbed will no longer be in the beam. Since the energy of the photon is proportional to the frequency, if we now pass the beam through a spectrometer, little or no light will be found at those frequencies corresponding to the energies of the absorbed photons. Thus in the spectrum we see dark lines, places where there is little energy, at the frequencies where photon absorption has occurred. In many cases, molecules produce groups of absorption lines so closely spaced that sections of the spectrum are black. Such groups are known as absorption bands.

Where does the energy absorbed by the molecule go? Some molecules remain in excited states; some collide with other molecules and transform the energy into motion, heating the sample; and some energy is reradiated by the sample. Since the direction of the reradiation is random, the photon is very unlikely to come out in the direction of travel of the original beam and therefore will never enter the spectrometer.

As you probably know, in the case of a simple atom like hydrogen, the energy states may be represented as electron orbits of increasing size about the nucleus. You may also know that in more complicated atoms, only the valence electrons, those from the outer orbit, play a role in the production of optical spectra. In the atomic case, the possible orbits are identified by the orbital quantum number n, with $n = 1$ being the lowest orbit. However, for a given n, orbits of different angular momenta are possible and each has a slightly different energy. The projection of the orbit in space also plays a role in determining the energy, as does the direction of the spin of the electron. All of these matters can be completely explained with quantum mechanics. An essential result of that explanation is that the various contributions to the energy of the system are quantized and can be described by expressions whose only variables are integers, the quantum numbers. Thus the particular state of a system is often expressed by writing the quantum numbers for that state.

In the molecular case, there are several contributions to the energy of the system. In addition to the contribution of the outer electrons, the molecule can both rotate and vibrate, and each of these is a quantized motion.

The energy levels for these modes may be found by considering the case of a diatomic molecule. The kinetic energy of rotation of a single particle about some point is

$$KE = \tfrac{1}{2} m\omega^2 r^2$$

The moment of inertia is

$$I = mr^2 \qquad \text{and} \qquad \therefore KE = I\omega^2/2$$

If we generalize to two particles, the form of the above is unaltered, provided that we introduce the reduced mass, defined by

$$\mu = m_1 m_2 / m_1 + m_2$$

Hence the case of two particles in mutual orbit is equivalent to the motion of a single particle of mass μ about a point. Then

$$I = \mu r^2 \qquad \text{and} \qquad L = I\omega$$

The energy of rotation ϵ_{rot} is quantized and given by

$$\epsilon_{rot} = \frac{h^2}{8\pi^2} \frac{1}{I} J(J+1)$$

These results can be generalized to more complicated systems. Nevertheless, the major result is clear. The rotational energy is quantized and the energy of a particular rotational state depends on a variable integer, the rotational quantum number, J.

Photons can be absorbed only if their energy equals the energy difference between two rotational states. It can be shown, either theoretically or empirically, that changes in the rotational quantum number are such that

$$\Delta J = \pm 1$$

Therefore, the energy difference between a lower state J_1 and upper state J_2 can be written

$$\Delta E_{rot} = \epsilon_{rot}^{(2)} - \epsilon_{rot}^{(1)} = \frac{h^2}{8\pi^2} \frac{1}{I} [(J_1 + 1)(J_1 + 2) - J_1(J_1 + 1)]$$

and since

$$J_2 = J_1 + 1$$

we have

$$\Delta E_{rot} = \frac{h^2}{8\pi^2} \frac{1}{I} (J_1{}^2 + 2J_1 + 2 - J_1{}^2 - 1) = \frac{h^2}{8\pi^2} \frac{1}{I} (2J_1 - 1)$$

For the vibrational case, the problem is a bit more subtle. The frequency of vibration will depend on the forces between the atoms in the molecule. These may be quite complex. In the diatomic case, a good beginning approximation is to assume a Hooke's law force: the force on a particular atom acts to restore a displaced atom to its equilibrium position and increases linearly with the increasing displacement of the atom from equilibrium. Thus

$$f = -kx$$

Since

$$f = mx$$

we have

$$-kx/m = \ddot{x}$$

The solution of this differential equation gives the frequency of vibration as

$$\nu = \frac{1}{2\pi} \sqrt{k/m}$$

We generalize by introducing the reduced mass; as before, this reduces the diatonic system to an equivalent single particle case. The quantized expression is then

$$\epsilon_{vib} = \frac{h}{2\pi} \sqrt{k/\mu} \; (v + \tfrac{1}{2})$$

where v is the vibrational quantum number, which can change so that

$$\Delta v = \pm 1$$

Thus

$$\Delta E_{\text{vib}} = \epsilon_{\text{vib}}^{(2)} - \epsilon_{\text{vib}}^{(1)} = \frac{h}{2\pi} \sqrt{k/\mu} \; [v^{(2)} - v^{(1)}]$$

In most biophysical situations, the rotational energy changes are the most difficult to observe because we are usually studying molecules in solution, and the collisions tend to damp out the small energy changes associated with rotational transitions.

These simple pictures, although only approximately correct, are useful both because they give one a feel for the quantities involved and, more significantly, because the particular small groups that compose a macromolecule often appear, from a spectroscopic viewpoint, to be quasi-independent of the rest of the molecule. This means that the presence of absorption features at certain frequencies, which will vary over a small range with the degree of isolation of the group, is characteristic of particular groups of atoms.

Since there are so many small perterbations, the spectra of macromolecules, which are observed in liquid or solid form, are characterized by broad absorption bands, and it is possible to describe the absorption without regard to molecular details. Consider the diagram in Figure 2.6.

$$I_e = I - dI$$

where I_e = number emergent. We assume that the molecules are independent and randomly arranged, and that they are illuminated by monochromatic light. Some photons in the incident beam will be absorbed. Therefore, the emerging beam is

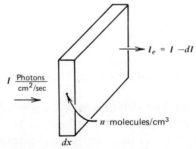

Figure 2.6 A monochromatic beam of intensity I is incident on a sample whose thickness is dx and has n absorbers per cm³. The emergent beam has an intensity I_e, which is less than I by some amount dI.

reduced by some amount $-dI$. Thus the probability that a photon is absorbed is proportional to

$$- dI/I$$

From the absorber's viewpoint, the probability of absorbing a photon is

$$\begin{array}{ccc} \text{Probability of} & \text{absorber} & \\ \text{molecular absorption} \times & \text{density} & \times \text{thickness} \end{array}$$

Thus we have

$$\text{Cross section, } \epsilon \times n \times dx$$

or

$$-\frac{dI}{I} = n\epsilon\, dx$$

$$\int_{I_0}^{I} dI/I = - \int_{0}^{x} n\epsilon\, dx = -n\epsilon \int_{0}^{x} dx = -n\epsilon x$$

so

$$I/I_0 = e^{-n\epsilon x}$$

This is the Beer-Lambert absorption law.

It is important to remember that the above result depends on the assumption of monochromatic illumination. If we illuminate a sample with a wide range of λ's, the probability for absorption of some of these λ's will be zero and the above result will be invalid.

The results we have obtained so far are useful mainly in finding out information about the molecule itself, especially when it is more or less isolated or in a known solution. In many cases, however, we are interested in changes produced in the state of the molecule when it is subjected to a particular environment or treatment. For example, the spectrum of a protein is not exactly the same as the spectrum of its constitutent amino acids; similarly, changes in the folding of a protein, as in denaturation, produce changes in spectra. These changes, which often are conforma-

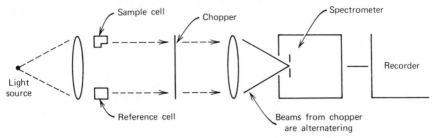

Figure 2.7 The above drawing shows the arrangement of the components of a dual-beam spectrometer.

tional in nature, are observed by difference or dual beam spectroscopy, as illustrated in Figure 2.7. One of the most impressive example of such a change in a spectrum occurs when DNA is denaturated by heating (see Chapter 5).

A common aspect of biopolymers is the presence of helical structure. This is an important feature from the standpoint of spectroscopy because this assymetry of structure creates the possibility for an interaction between light and the biopolymer that depends on the polarization of the incident light.

Polarized light is light in which the **E** vectors are not uniformly distributed at right angles to the direction of propogation. In plane polarized light, the **E** vectors oscillate in a plane; in circularly polarized light, they rotate about a line marking the direction of propogation. Plane polarization is equivalent to the sum of two circularly polarized components that rotate in opposite directions, while elliptically polarized light is equivalent to this sum when the components are of unequal amplitude.

Two particular effects may be observed. In the first, called circular dichroism, we find some molecules in which the absorption is stronger for one component than the other. The second possibility is that the plane of polarization is rotated from its incident value when plane polarized light is shined through the sample. This is known as optical rotation. The variation of this effect with wavelength is optical rotary dispersion, and it is possible to derive one result from the other. The principal application of these effects in biophysics is to the study of helicity in biopolymers.

Such structural features violate one of the assumptions made in deriving the Lambert-Beer law. The consequences of this can be shown as follows: Consider a group oriented with respect to a long chain polymer that is part of a collection of aligned polymers. The situation is illustrated in Figure 2.8. The incident light is of course plane polarized, so

$$E = E_y + E_x$$

where

$$E_x = E \cos \theta \qquad \text{and} \qquad E_y = E \sin \theta$$

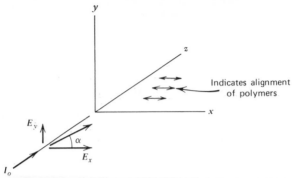

Figure 2.8 A plane-polarized light wave, oriented at an angle α, is incident on a set of aligned polymers.

Intensity is proportional to the square of the amplitude of the wave:

$$I_x = I_0 \cos^2 \theta \qquad \text{and} \qquad I_y = I_0 \sin^2 \theta$$

Now let ϵ_y and ϵ_x be the cross sections for different components. Then

$$-dI_x/I_x = n\epsilon_x \, dz \qquad \text{and} \qquad -dI_y/I_y = n\epsilon_y \, dz$$

However, the value of $\epsilon_y = 0$ because the electric field E_y in our situation is at right angles to the direction of possible electron motion; therefore, the electric field cannot interact and no absorption can occur. Thus

$$-dI_x/I_x = n\epsilon_x \, dz \qquad \text{but} \qquad -dI_y/I_y = 0$$

So

$$I_x = I_{x0} e^{-n\epsilon_x z} \qquad \text{and} \qquad I_y = I_{y0}$$

We measure the total transmitted intensity:

$$I_x + I_y = I_{x0} e^{-n\epsilon z} + I_{y0}$$

which is obviously not equivalent to

$$I = I_0 e^{-n\epsilon z}$$

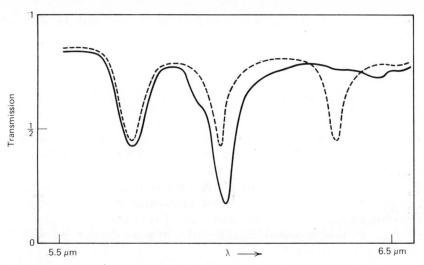

Figure 2.9 The absorption spectra of a film made of α-helix polymers which have been aligned by stretching the sheet. The dotted curve shows the absorption ⊥ to the polymer axis; the solid line, ‖ to it. (Tsuboi, J.P.S. *59*, 1962)

The above case is deliberately extreme. However, groups can be limited so that the orientation is decidely nonrandom, producing, as above, deviations from the Lambert-Beer law. The spectra in Figure 2.9 show the effect clearly.

It is also possible to get information about macromolecules through observations of their light scattering properties; the experimental arrangement is shown in Figure 2.10. Light scattering was initially applied to the study of macromolecules by Debye, who used that method, in 1943 to 1944, to get molecular weights, sizes, and shapes from measurements of I as $f(\theta)$. In treating the light scattering problem, it is convenient to consider two separate cases, depending on whether or not the macromolecule is

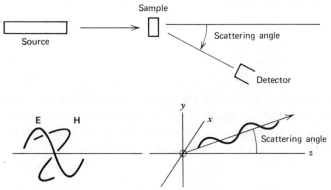

Figure 2.10 This figure shows the schematic arrangement of the components for a light scattering measurement.

larger than the wavelength λ of light being scattered. We begin with the case in which the size of the macromolecule is $<\lambda$, and we deliberately choose a wavelength far from any absorption band. The electric field of the incident radiation is described by

$$\mathbf{E} = \mathbf{E}_0 \cos 2\pi\nu t$$

As in the case of X rays, this $\mathbf{E}$ exerts a force on the electrons. However, the frequency of visible light is very much lower than that of X rays, and the energy of the photons in the incident beam is correspondingly reduced. The electrons are shaken back and forth, which means that the distribution of the electronic charge varies periodically. Therefore, since the charge distribution in the macromolecule is asymmetric, the center of this distribution is not a fixed point.

When we have opposite charges separated by some distance, we say a dipole moment is present, given by

$$\mathbf{p} = q\mathbf{x}$$

where q is the charge and x the separation between them. The effect of the electric field on the macromolecule is to produce a dipole moment by displacing the electron distribution; such a molecule is said to be polarized. In our case, the dipole moment will oscillate with the field:

$$\mathbf{p} = \alpha\mathbf{E} = \alpha\mathbf{E}_0 \cos 2\pi\nu t$$

where α is the polarizability, the proportionality constant between the dipole moment and the electric field that produced it. It can then be shown that

$$\frac{I \text{ scattered}}{I \text{ incident}} \sim \frac{\sin^2 \theta}{r^2 \lambda^4}$$

that is, the intensity falls off with the square of the distance from the scatterer, the fourth power of λ, and the square of the angle between the incident radiation and the dipole moment. It is usually more convenient to deal with unpolarized light and the angle between the incident and scattered radiation. Then

$$\frac{I_s}{I_i} = \frac{8\pi^2 \alpha^2}{r^2 \lambda^4} (1 + \cos^2 \theta)$$

Normally, we observe scattering from a macromolecular solution. Therefore, there are two sources of scattering, the molecules of the solvent and the macromolecules themselves. Thus we want to deal with the excess of scattering over that observed in the pure solvent.

In electromagnetic theory, a connection between the index of refraction and the polarizability for a transparent dialectric is derived; this result is

$$\frac{n^2 - 1}{n^2 + 2} = \tfrac{4}{3}\pi N\alpha \qquad \text{where} \qquad N = \frac{\text{No. of particles}}{\text{cm}^2}$$

or, since $n \approx 1.3$ in the case of aqueous solutions,

$$n^2 - 1 \approx 5\pi N\alpha$$

Suppose we measure the excess refractive index, the difference between n_{solu} and n_{solv}. Then we can write

$$n^2_{\text{solu}} - n^2_{\text{solv}} = 5\pi N\alpha = n'^2 - n_0^{\,2}$$

where α is now the excess polarizability produced by the displacement of solvent by solute molecules. Then

$$(n' - n_0)(n' + n_0) = 5\pi N\alpha$$

so

$$\alpha = \frac{(n' + n_0)}{5\pi} \frac{(n' - n_0)}{c} \frac{c}{N}$$

Where c is the concentration in g/cm^3. We define

$$\frac{n' - n_0}{c} = \text{specific refractive index increment}$$

and note that

$$\frac{c}{N} = \frac{\text{Molecular weight}}{N_A} = \frac{M}{6.02 \times 10^{26}}$$

If the solution is relatively dilute, $n' + n_0 \simeq 2n_0$, and therefore

$$\frac{2n_0}{5\pi} \frac{dn}{dc} \frac{M}{N_A} = \alpha$$

This may now be substituted for α in the equation for the ratio I_s/I_i, giving

$$I_s/I_i = \frac{1.3n_0^2 (dn/dc)^2 M^2}{\lambda^4 r^2 N_A^2} (1 + \cos^2 \theta) \qquad \text{per particle}$$

For N particles/cm^3,

$$N = \frac{CN_A}{M}$$

and we have

$$I_s/I_i = \frac{1.3n_0^2}{\lambda^4 r^2 N_A} (dn/dc)^2 CM (1 + \cos^2 \theta)$$

Thus we can obtain from light scattering measurements the molecular weight of a macromolecule when its concentration in a solvent is known. The major assumption is the statement that the scattering of N molecules is N times the scattering of a single molecule. Simple formulas to correct the inaccuracy introduced by this assumption have been derived from thermodynamics by Debye. The use of these corrections complicates the experiment by requiring measurements to be made at different concentrations, but nothing else is altered.

In the preceeding treatment, we tacitly assumed that the molecule was small compared to λ when we assumed that the polarization occurred over the whole molecule simultaneously. If the molecule is in fact large, the value of E is not necessarily the same at a given t for the whole molecule. Thus we are led to consider the case in which the size of the macromolecule is $> \lambda$.

We can approximate a large macromolecule as a collection of dipoles. Let

$$P(\theta) = \frac{\text{Observed intensity at } \theta}{\text{Intensity expected if the molecule were small compared to } \lambda}$$

Then it can be shown that

$$P(\theta) = \frac{1}{N^2} \sum_{i,j} \frac{\sin R_{ij}h}{hR_{ij}}$$

where $\quad h = \frac{4\pi}{\lambda} \sin \tfrac{1}{2}\theta$

N = number of scattering centers $\simeq$ number of atoms

R_{ij} = separation of scattering centers $\simeq$ atomic separations

Furthermore, if $\theta > 0$, $P(\theta) < 1$; therefore, scattering for a large molecule is always less than for a small molecule of equal molecular weight. Now let us define the radius of gyration:

$$R_G \equiv \frac{1}{2N^2} \sum_{i,j}^{N} R_{ij}^2$$

This is the distance from a particular axis at which the entire mass could be concentrated so that mR_G^2 would equal the actual moment of inertia. Then, in approximation,

$$P(\theta) = 1 - \frac{h^2 R_G^2}{3} = 1 - \frac{16R_G^2}{3\lambda^2} \pi^2 \sin^2 \frac{\theta}{2}$$

Therefore, for larger macromolecules, the angular dependence of the light scattering can yield the shape of the molecule in the form of a value for the radius of gyration; some values of R_G are given in Table 2.1.

The development of lasers has greatly stimulated interest in light scattering investigations by making it possible to get much more detailed information than was previously the case. It is now possible to obtain excellent values for translational and rotational diffusion coefficients of molecules and macromolecules, as well as information on reaction kinetics. Light scattering is an especially attractive method for studying the distribution of macromolecular sizes in solution, since it does not change

Table 2.1

R_G	Particle
$\sqrt{3/5}\,R$	Sphere of radius R
$L/\sqrt{12}$	Rod of length L
$\sqrt{\overline{\phi^2}}/\sqrt{6}$	Random coil with end to end length = ϕ

the equilibrium between the various conformations in the solution. In one technique, developed by Benedek, the Brownian motion is studied by determining the diffusion constant, D, which is inversely proportional to the size of the macromolecule. If one allows monochromatic radiation, say from a laser, to be scattered by a solution of macromolecules, the sharp line is broadened by an amount proportional to D. The spectrum of scattered light is governed by the distribution of diffusion constants, which can be obtained by measuring the fluctuations in the intensity of the scattered light. These fluctuations are analyzed by the correlation function:

$$R(t) = A \sum_n \omega_n \exp-(\ D_n h^2 t)^2 = \int_0^\infty \omega(D) \exp-(Dh^2 t)^2\, dD$$

where A is an experimentally determined constant proportional to the amplitude of the scattering and ω_n is a weighting function for the nth conformational form, taken so that

$$\sum_n \omega_n = 1$$

where the sum is over the n possible conformations.

Then $$\int \omega(D)dD = 1$$

and $\omega(D)dD$ is the relative intensity of light scattered by a particle with a diffusion constant between D and $D + dD$. The correlation function is obtained from the experiment; the distribution of diffusion constants in the solution, and thus the distribution of macromolecular sizes, is obtained by interpolating the correlation function into a set of theoretical curves derived for various distributions.

So far we have considered the optical region of the electromagnetic spectrum. Let us close this section by briefly considering a spectroscopic technique for another region.

Radio frequency spectroscopy is in a sense no different from optical spectroscopy: Electromagnetic radiation is passed through a sample and the frequencies at which absorption occurs are determined. However, the absorptions are now due to magnetic dipole changes in the sample. In the case of nuclear magnetic resonance (NMR), one observes changes in the nuclear magnetic moment, while in the case of electron spin resonance (ESR), it is the electron magnetic moment that changes. Since the electron magnetic moment is about 1000 times larger than the nuclear magnetic moment, the two techniques are substantially different with respect to apparatus, problems, and results. We consider only ESR.

Since ESR studies electron magnetic moment changes, it follows that the only systems that we can investigate are those in which the electron magnetic moment is not zero. Thus, all diamagnetic materials are excluded; only paramagnetic substances, that is, those with an unpaired electron can be used. Two general sources for paramagnetic substances are immediately obvious: ions from the transition elements and free radicals and their ions, in which the valence electron is the unpaired electron.

In order to observe resonance spectra, the sample is placed in a microwave cavity, at a point corresponding to H_{max} for the microwave. The sample cavity is placed in a magnetic field so that H_{mag} is $\perp$ to $H_{max}^{microwave}$. H_{mag} is then varied until a resonance occurs. The justification for this technique can be seen from the following argument. Suppose we have an isolated electron in a magnetic field. The electron can line up with its spin in the direction of the field or opposite of it. The energies of these two states are slightly different. If we define the Bohr magneton:

$$\mu_0 = eh/4\pi mc$$

then the energies of the two states are $+\mu_0 H_{mag}$ and $-\mu_0 H_{mag}$. Therefore, the energy absorbed in a transition should be

$$h\nu = \mu_0 H_{mag} - (-\mu_0 H_{mag}) = 2\mu_0 H_{mag}$$

Hence, by varying H_{mag}, we can find the value required to produce an absorption of radiation at the frequency ν. From this, it is easy to see that ESR is a form of microwave spectroscopy applied to paramagnetic materials, usually at a fixed frequency.

In actual situations, the electron is not an isolated particle in the magnetic field, but is affected by interactions with other electrons and nuclei. It is convenient to introduce a purely empirical term, the g-factor, which is a measure of the shift produced in the resonance by those interactions. The g-factor is defined by

$$h\nu = g\mu_0 H_{mag}$$

where, in the case of a free electron,

$$g = 2.0023 \pm 1\%$$

From the above, it follows that the ESR spectrum of a material could be characterized by some shift in the resonance line away from the totally free-electron position. This can be expressed by giving the appropriate g-value. However, the problems of analysis of such results are formidable, owing to the complexity of the interactions. For example, it is usually not possible to identify a particular free radical solely from the g-value determination. Such radicals can be identified through a study of the resonance line splitting, usually called hyperfine splitting. This splitting is due to the interaction between the electron magnetic moment and the magnetic moments of the various nearby nuclei. For example, suppose the electron interacts with a nearby proton. The proton has a spin of $\frac{1}{2}$ and therefore can align with or against H_{mag}; clearly, half the electrons (unpaired of course) "see" a field $> H_{mag}$ and the other half (see a field $< H_{mag}$. The line is thus split into two equal components. This, of course, is the simplest situation; as the spin of the nuclei involved increases, so does the number of possible nuclear orientations and hence the number of possible components. Finally, there are a variety of interactions that can affect the width of the line. Usually not much information can be extracted from line width determinations; wide lines are mainly an impediment to the observation of splitting.

Both ESR and NMR have been applied to a variety of problems in biophysics, ranging from studies of radiation effects on materials, through the determination of diffusion constants in solutions, to the study of dynamic processes in cell components.

ELECTROPHORESIS

An important property of proteins is that they may be charged; this leads us to investigate their behavior in electric fields. There are two possible ways in which an electric field can affect a macromolecule:

    1. Motion will occur because the molecule has a net charge:

$$F = zeE$$

The molecule is subject to friction as it moves, $\therefore$ a friction factor is introduced:

$$f = zeE/v$$

This leads us to define the mobility:

$$U = v/E = Ze/f$$

From Stokes' law,

$$U = Ze/6\pi\eta R$$

for spherical or quasi-spherical particles of radius R.

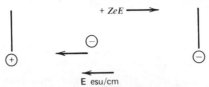

Figure 2.11 Electrophoretic motion.

2. Macromolecular orientation will occur due to the asymmetry of charge distribution. Such asymmetry may be intrinsic, or the molecule may be polarized by the action of the impressed electric field.

One of the most important practical consequences of motion in an electric field is electrophoretic separation or electrophoresis. Since macromolecules such as proteins have a variety of shapes, and therefore a range of frictional drags in a medium, as well as a range of net charges that are usually a function of the pH of the medium, it follows that different proteins will move at different rates through a medium. Thus they will become separated. The process is illustrated in Figure 2.11.

VISCOSITY MEASUREMENTS

The resistance of a fluid to a completely free flow is its viscosity. This is expressed quantitatively by the experimental arrangement shown in Figure 2.12. We define shear strain as $dx/dy = \Sigma$ and shear stress as F_x/A_{plate}. For so-called Newtonian fluids,

$$\eta \dot{\Sigma} = F_x/A$$

where η is the coefficient of viscosity. Viscosity is also proportional to the rate of energy dissipation in the flow. If we express the shear rate as

$$v/\Delta y$$

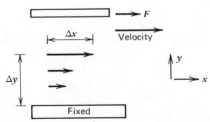

Figure 2.12 The viscosity of a fluid may be defined with the above apparatus, in which the fluid is placed between the plates, and the force required to move one plate with respect to the other is measured.

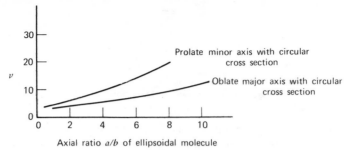

Figure 2.13 ν values from viscosity measurements.

then

$$(F/A\,\Delta y)\dot{x} = \eta(v/\Delta y)^2$$

whence

$$dE/dt = \eta(v/\Delta y)^2$$

If we now measure the viscosity of a pure solvent, we find that the viscosity of a solution of macromolecules in this solvent is greater. Therefore,

$$(dE/dt)_s = (dE/dt)_0\,(1 - \nu\phi)^{-1}$$

where ϕ is the fraction of the solution volume occupied by the macromolecules and ν is a proportionality factor. To a good approximation,

$$(dE/dt)_s \doteq (dE/dt)_0\,(1 + \nu\phi)$$
$$\therefore \eta_s/\eta_0 \doteq 1 + \nu\phi$$

Values of ν for different shapes of particles are shown in Figure 2.13.

chap. 3

The Elemen-tary Physics Of Macromol-ecules

Although some of the properties of macromolecules in living matter are probably strongly influenced as a result of their association in complex structural arrangements, we believe that at least some of the important properties of these molecules will be preserved even when the molecules are studied in isolation by the methods previously described. For example, we know that protein coiling depends on temperature, among other things, and a study of helicity in proteins provides an understanding of protein denaturation, the order-disorder changes that can occur in cells. Here, of course, we could raise the *in vivo, in vitro* question; it is very hard to be absolutely certain that experiments on components always give results that can be extrapolated to the properties of components in cells. This difficulty should not be overemphasized. We have many examples, such as enzyme rate studies and DNA replication experiments, that yield important insights into the corresponding cellular processes. Furthermore, any difficulty with extending *in vitro* results to *in vivo* conditions certainly has nothing to do with any failure of physics or chemistry. If we can match cellular complexities — and we will — we will obtain cellular results.

ATOMIC AND MOLECULAR FORCES

We now ask: What are the forces that shape these macromolecules not only in themselves but also in the assemblies that form the structural basis for living matter? To answer this question, we have to investigate the forces between atoms that lead in the first place to the formation of molecules and then the forces that act between these molecules.

In this discussion, it is convenient to characterize the electromagnetic forces as strong or weak, depending on whether the binding energy due to the force is large or small compared to the thermal energy, kT, at a temperature appropriate to biological material.

Furthermore, although we discuss structures in terms of the forces that act, it is important to realize that in experiments we measure the interaction energy produced by the action of the force. The connection between this interaction, or potential energy, and the appropriate force is quite specific; it is

$$F = -\text{grad } V,$$

where V is the potential energy (see Appendix 2).

There are three kinds of strong electromagnetic forces that can exist between atoms; the three types of bonds formed by these three forces are (1) ionic, (2) covalent, and (3) resonant or exchange.

> **1.** In simplest terms, the ionic bond, of which NaCl is the classic example, is formed when an atom with a low-ionization potential has its outer electron "captured" by an atom with high-electron affinity. This leaves the first atom as a whole positive, and the second negative; the Coulomb attraction

between the two holds them together. There is a reasonably well-defined equilibrium separation since, if the atoms come close enough for their electron clouds to overlap, Coulomb repulsion between the positively charged nuclei will occur.

2. The covalent bond is usually said to be formed by the "sharing of electrons." The understanding of this force depends on a quantum mechanical explanation that was first provided by Heitler and London in 1927. The covalent bond is essentially a consequence of the uncertainty principle, which states that

$$\Delta x \, \Delta P_x \geqslant \frac{h}{2\pi}$$

where Δx is the uncertainty in position and ΔP_x is the uncertainty in momentum. As two atoms approach each other, the outer electrons, which are not at localized points in space but instead must be thought of as a "smear" of charge around the atom, cease to be assignable to a particular atom. This means that the uncertainty in the electron's position is now greater than it was when the two atoms were separate. Therefore, the uncertainty in the momentum, and thus the minimum possible momentum, decreases. Since momentum and energy are related,

$$\tfrac{1}{2}mv^2 = E \qquad mv = P \qquad \text{so} \qquad E = P^2/2m$$

the energy of the system also decreases. But a decrease in the energy of the system is an increase in the binding, and therefore the atoms are held together. Again, there is an equilibrium separation, since if the atoms come too close, the volume accessible to the electron decreases, which increases the possible momentum and therefore the energy. Of course, the above is a very qualitative explanation indeed, and there are several critical points that are really understandable only by the full mathematical analysis; nevertheless, the above is roughly correct. Since each atom usually contributes one electron, the bond is known as *co*valent. Complex atoms

Table 3.1 Covalent Bond Energies

Bond	Energy (eV)
H–H	4.4
C–C	2.6
C=C	4.4
C≡C	5.4
Peptide bond	3.0
C–H	3.8

can, of course, share more than one pair of electrons, thereby increasing the strength of the bond. However, as Table 3.1 shows, such an increase is not linear with the number of pairs shared.

3. The third type of bond, the resonance or exchange bond, is essentially a more complex version of the previous explanation. In some systems, for example, benzene, several different electron configurations have essentially identical probabilities. Therefore, a particular molecule cannot be thought of as having a specific configuration. The effect of this is to increase the volume available to the electrons and, as before, to increase the binding energy. Again, a precise explanation requires a quantum mechanical treatment. These predictions are borne out by direct measurement of the interatomic spacing using X-ray diffraction. For example, the separation of carbon atoms in the absence of exchange forces is .154 nm; in benzene, where resonance occurs, the separation is .139 nm: the increase in the binding energy due to the resonance decreases the separation. Exchange forces also occur in the peptide bonds of proteins; the binding energy in that case increases by about 1 eV and the C–N separation is decreased by .016 nm.

We now turn to the weak forces that can act not only between atoms but also between molecules. These forces are as follows:

1. *Dipole–dipole interactions.* The center of the distribution of positive charges in a molecule is not necessarily coincident with the center of the distribution of the negative charges. If the separation of these centers is ℓ, then the dipole moment p is defined as

$$p \equiv q\ell$$

If such a dipole is subject to an electric field, a force acts on it, which if unopposed, will align the dipole with the direction of the electric field (see Figure 3.1). The work required to produce this alignment is

$$d\omega = pE \sin \theta \, d\theta$$

Thus if two dipoles are brought together, there will be a force between them owing to the action of the electric field produced by one on the other. The dipole will turn so that, relative to the first, it has the lowest possible energy. The average electric field due to dipole A is

$$E_A = P_A / 4\pi r^3 \epsilon$$

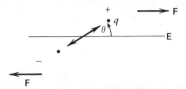

Figure 3.1 A dipole aligning with an electric field.

where ϵ is the dielectric constant of the medium. Since $d\omega = pE \sin \theta \, d\theta = F \, dx$ and $F = -\text{grad } V$, we have $V = -P_B E_A \cos \theta = -P_B E_A$ for the lowest energy. Therefore

$$V = -P_B E_A = -P_B P_A / 4\pi r^3 \epsilon$$

The orientation of a set of dipoles will not be perfect; not all will have $\theta = 0$. This is because they are subject to random, thermal agitation. The distribution of θ can be found from the Boltzmann equation, a general law that states that the ratio of the number of systems with one energy to the number with another goes as the exponential of the energy difference between the states, divided by kT:

$$n_1/n_2 \sim e^{-\Delta E/kT}$$

In our case, the energy difference is V, and V is a function of θ. We obtain an expression for $V(\theta)$ by

$$V_\theta = \int_0^\theta d\omega$$

$$= pE \int \sin \theta \, d\theta$$

$$= pE(1 - \cos \theta)$$

$$\therefore \frac{n_\theta}{n_o} = e^{-pE(1-\cos\theta)/kT}$$

$$= e^{-pE/kT} e^{pE \cos \theta / kT}$$

The average dipole moment $\bar{p}_B$ can be found if

$$\bar{P}_B E_A \ll kT$$

to be

$$\bar{P}_B = 1/3 \; E_A P_B / kT$$

So in the normal situation, when strong thermal agitation is present,

$$V = -\bar{p}_B E_A = -\tfrac{1}{3} E_A{}^2 p_B{}^2 / kT$$

$$= -\tfrac{1}{3} p_A{}^2 / (4\pi\epsilon)^2 \times p_B{}^2 / kT \times \frac{1}{r^6}$$

The basic dipole—dipole interaction can be extended to several other cases.

2. A dipole moment may be induced by the presence of a molecule with a permanent dipole moment. In this case, the induced dipole moment is dependent on the strength of the electric field due to the molecule with a permanent dipole moment, and therefore

$$p_B = \alpha E_A$$

where α, the proportionality constant, is called the polarizability. Then, as before,

$$V = -\alpha P_A{}^2 / (4\pi\epsilon)^2 r^6$$

3. There are many cases of symmetrical molecules, which are therefore without dipole moments, but nevertheless attract each other; for example, H_2 attracts H_2. This attraction is produced by fluctuations in the charge distribution. These fluctuations produce a transient dipole that acts to induce dipole moments in adjacent molecules. Nevertheless, the interaction energy still depends on $1/r^6$. In fact, this transient-induced dipole interaction is the dominant force both for small permanent dipole molecules and molecules in solution.

4. The most important weak interaction in biological molecules produces the hydrogen bond. Allen has shown that the hydrogen bond can be analyzed with only three quantities: the dipole moment, p; the bond length with the electron donor, R; and the ionization potential, ΔI, taken with respect to the noble gas atom in that row of the periodic table that contains the electron donor. The energy of the bond is given by

$$E = \text{constant} \cdot \frac{p(\Delta I)}{R}$$

This interaction occurs by covalent bonding between hydrogen and oxygen, or nitrogen. These groups have strong dipole moments, and there is a strong, positive charge near the proton because the electron has been drawn toward the other nucleus. This means that other negative groups, either in other molecules or in the same molecule, may be bound by Coulomb interaction with the "semiexposed" proton of the hydrogen atom, with a binding energy of about 5% of that of a true covalent bond.

5. The final interaction is known as charge-exchange. It occurs because many macromolecules are charged when in solution. In a manner similar to the transient dipole interaction, charge fluctuation occurs and this leads to a fluctuating dipole moment. This produces a short-range, weak interaction.

BEHAVIOR OF MACROMOLECULES

Now consider a linear polymer, say a protein composed of a chain of amino acids. If each amino acid is free to rotate about a bond, the protein will coil randomly. A collection of such coils in solution will show a variety of shapes. We can calculate, by the Monte Carlo or random walk approach, the average separation of the ends of a coiled strand formed from N equal segments. Let us take the end of the first segment as the origin. Then the distance from the origin, in terms of a segment length, of the end of a chain of segments will be

$$m = N_1 - N_2$$

where N_1 is the number of segments pointing away from the origin when added, and N_2 is the number pointing toward the origin.

$$N = 2N_2 + m$$
$$N_2 = N/2 - m/2 \qquad \text{and} \qquad N_1 = N/2 + m/2$$

The probability of a given m, P_m, is then

$$P_m = (\tfrac{1}{2})^N \frac{N!}{N_1! \, N_2!} = \frac{(\tfrac{1}{2})^N N!}{(N/2 + m/2)!(N/2 - m/2)!}$$

To obtain a more useful expression, use Stirling's approximation:

$$N! = (2\pi N)^{\tfrac{1}{2}} (N/e)^N$$

whence

$$\log N! = (N + \tfrac{1}{2}) \log N - N + \tfrac{1}{2} \log 2\pi$$

After some algebra and the use of the relation:

$$\log(1 + x) \doteq x - \frac{x^2}{2}$$

it can be shown that

$$\log P_m = -\tfrac{1}{2} \log 2\pi + \log 2 - \tfrac{1}{2} \log N - m^2/2N$$

or

$$\log P_m \doteq \log \frac{2}{\sqrt{2\pi N}} - \frac{m^2}{2N}$$

So

$$P_m = \frac{2}{\sqrt{2\pi N}} e^{-m/2N}$$

Now let l be the length of a segment and ζ the total displacement. Then

$$P_m = \sqrt{\frac{2}{\pi N}} e^{-\zeta^2/2l^2 N}$$

The probability of a particle finding itself a distance from $(0,0)$ is

$$P_\zeta \, d\zeta = P_m \, d\zeta/2l$$

$$P_\zeta = P_m/2l = \frac{1}{\sqrt{2\pi l^2 N}} e^{-\zeta^2/2l^2 N}$$

Since $\lambda^2 = l^2 N$,

$$P_\zeta = \frac{1}{\lambda\sqrt{2\pi}} e^{-\zeta^2/2\lambda^2}$$

The mean square displacement from $(0, 0)$ is

$$\overline{\zeta^2} = \frac{\int_0^\infty \zeta^2 P_\zeta \, d\zeta}{\int_0^\infty P_\zeta \, d\zeta} = \lambda^2$$

and so the average separation of the ends will then be given by

$$\zeta^2 = l^2 N$$

The radius of gyration of such a coil is found from

$$\frac{l^2 N}{6} = \overline{R_G^2}$$

As you already know, R_G may be determined from light scattering experiments.

The random walk argument gives the probability for various coil sizes; of course these arise from various coil shapes. A variation in coil shapes suggests that there could be a variation in the degree of ordering from coil to coil. The concept of order is closely connected to thermodynamics (see Appendix 2), and in fact it is found that useful insights into macromolecular properties can be obtained through an appeal to thermodynamic principles.

In a thermodynamic analysis it is convenient to distinguish between properties associated with molar quantities and those associated with the specific quantity of material; partial molar and partial specific quantities are written

$$\overline{X_i} \equiv \partial X/\partial n_i \qquad \text{where } n_i \text{ is the number of moles}$$
$$x_i \equiv \partial x/\partial g_i \qquad \text{where } g_i \text{ is the number of grams}$$

An essential quantity for describing macromolecular solutions is the partial molar-free energy G_i, which in this special case is called the chemical potential:

$$\overline{G_i} \equiv \mu_i \equiv (\partial G/\partial n_i)$$

The total free energy of the solution is

$$G = \sum_{i=1}^{N} n_i \mu_i$$

Now

$$dG = -S\,dT + V\,dP + \sum_i \mu_i\,dn_i$$

$$= \sum_i \mu_i\,dn,$$

if T and P are constant. Since we have

$$G = \sum_i n_i \mu_i$$

we get

$$dG = \sum_i \mu_i\,dn_i + \sum_i n_i\,d\mu_i$$

Therefore

$$\sum n_i\,d\mu_i = 0$$

and this means that variations in μ_i are not independent. Solutions of macromolecules are said to be ideal if:

1. There is no difference in interaction energy between solute and solvent. The enthalpy:

$$H = E + PV$$

where E is the internal energy, does not change upon mixing: $\Delta H_m = 0$.

2. The entropy change is due to only the change in randomness produced by mixing.

$$\Delta S_m = -R \sum_{i=1}^{n} n_i \ln X_i$$

where the X_i's are the mole fractions.

Therefore, the free energy of mixing is

$$\Delta G_m = \Delta H_m - T\Delta S_m = RT\Sigma n_i \ln X_i$$

In general, the free energy of mixing is

$$\Delta G_m = G_{solution} - \sum_i G_{i(\text{components in pure form})}$$

Therefore

$$\Delta G_m = \sum_i n_i\mu_i - \Sigma n_i\mu_i{}^0$$

so

$$n_i(\mu_i - \mu_i{}^0) = RT\sum_i n_i \ln X_i$$

and, since each component is independent,

$$\mu_i - \mu_i{}^0 = RT \ln X_i$$

This can also be written

$$\mu_i = \mu_i{}^0 + RT \ln c_i$$

where c_i is the concentration (grams/liter), since in most cases solute concentrations are low.

We don't expect macromolecular solutions to be ideal, just as real gases are not ideal. As you probably know, deviations from the ideal gas can be accommodated by using equations of state with the form

$$pV = nRT + f(T)$$

In a similar way, nonideal macromolecular solutions can be described by writing

$$\mu_i = \mu_i{}^0 + RT \ln c_i y_i$$

where

$$y_i = f(T, P, c_i\text{'s}) \qquad \text{and} \qquad \lim_{c \to 0} y_i = 1$$

Then a virial equation for the potential can be written:

$$\mu_i - \mu_i{}^0 \simeq - \frac{RTVc_2}{M} - \text{constant} \times c_2^2$$

where V is the molar volume of pure solvent, M is the molecular weight, and c_2 is the concentration of the solute.

One of the most important macromolecular solution phenomena that can be treated by the above method is osmotic pressure. The experimental situation is shown in Figure 3.2. For equilibrium,

$$\mu_i{}^\alpha = \mu_i{}^\beta$$

therefore

$$\mu_i - \mu_i{}^0 < 0$$

$\therefore$ Solvent must move from $\beta \to \alpha$

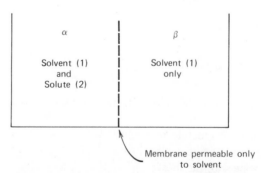

Figure 3.2 An arrangement for demonstrating osmosis. The chamber on the left, indicated by α, contains a solute dissolved in a solvent. The other chamber, β, contains only solvent.

and consequently equilibrium cannot be obtained without an additional force. This force is pressure. Let the pressure in β be P_0 and in α, $P_0 + \pi$. Then

$$(\partial G/\partial P)_T = V \qquad (\partial \mu_1/\partial P)_T = \overline{V_1} \qquad \mu_1{}^\beta = \mu_1{}^0$$

and

$$\mu_1{}^\alpha = \mu_1{}^0 - RTV_1{}^0(c_2/M_2 + Bc_2{}^2 + \ldots) + V_1{}^0\pi$$

so

$$\pi = RT(c_2/M_2 + Bc_2{}^2 + \ldots) \doteq RTc_2/M_2$$

Thus we have not only an explanation for the phenomenon of osmosis but also a method for determining the molecular weight of the solute.

Let us return for a moment to the macromolecule as a random coil. We know the simple random walk calculation will not be completely accurate because the macromolecular segments are not completely free to rotate about each bond. We know this because we already know forces that will restrict this motion: hydrogen bonds; the various dipole or van der Waals forces; and electrical effects, such as interactions between the macromolecules and the solvent molecules. The thermodynamic discussion tells us about energies and molecular weights; we now look for information on the structure in solution.

The first point is that the conformation is concentration dependent. Suppose ω is the number of possibilities for a segment. Then, if there are N segments, the number of conformations for the chain is ω^N. Then for n independent chains, the total number of conformations is

$$Q = (\omega^N)^n = \omega^{Nn}$$

If the volume fraction of the macromolecule is v, the number of conformations becomes reduced by e^{-vN}. Clearly, if $v = 1$, the solution is crystallized. Since ω is the number of possible segmental conformations, it is also a measure of the flexibility of the molecule. For the cases with $v \rightarrow 1$,

$$Q \doteq (\omega/e)^{Nn}$$

We can now begin to see how the previous matters can be brought together in order to understand the structure of proteins in solution. The strong forces, principally the covalent bonds, determine the primary structure of the protein. The secondary structure, helix or chain, is determined by the weak forces. When the protein is in solution, competition develops between the hydrogen bonds in the protein, which link NH–OC groups, and the hydrogen bonds that could form between these groups and water molecules, that is NH–OH$_2$ and CO–H$_2$O. Matters are more serious if heat or pH changes enter the picture, or if other components also compete with peptide groups in the formation of hydrogen bonds. If these various competitors are successful, the protein loses its biological activity even though the peptide bonds are not broken and the primary structure is unaltered. Such a protein is said to be denatured. By studying the forces that destroy secondary and tertiary structure, we get information on the forces that produce the structure. The denaturation process is accompanied by a free-energy change, ΔF, whose components may be summarized as follows:

$$F = \Delta F_1 \text{ from unwinding of } \alpha\text{-helix}$$
$$+ \Delta F_2 \text{ from breaking H-bonds in side-chain residues}$$
$$+ \Delta F_3 \text{ from hydrophobic interactions}$$
$$+ \Delta F_4 \text{ from electrostatic force changes}$$
$$+ \Delta F_5 \text{ from solvent effects}$$
$$+ \Delta F_6 \text{ from any crystallization that may be present}$$

All of this makes one wonder how the protein structure ever maintains itself in solution. The answer seems to be through hydrophobic interactions. The amino acids that make up the protein are in some cases charged (polar) and in other cases not. A contact between a nonpolar group and a water molecule is an increase in free energy and therefore unfavorable. Thus the nonpolar amino acid residues tend to be in contact with each other and not with the water molecules, while the polar residues face the water; as a result the protein folds up.

Fisher has given a simple argument to show the correctness of the above. Suppose we have a protein containing p polar residues and n nonpolar residues; n will be "inside" and p will be "outside." Let us imagine that the layer of p residues has a uniform thickness d. The volume of the outer layer is

$$V_e = \tfrac{4}{3} \pi [r^3 - (r - d)^3]$$

where r is the radius of the globule of protein. The internal volume is

$$V_i = \tfrac{4}{3} \pi [(r - d)^3]$$

Now

$$V_{total} = V_i + V_e$$

Thus

$$\frac{p}{n} = \frac{V_e}{V_i} = \frac{r^3}{(r-d)^3} - 1$$

So

$$V_e = V_t \frac{p}{p+1}$$

and

$$V_e = Ad$$

where A is equal to the surface area of the globule. Since Fisher uses $d \approx .4$ nm,

$$p = \frac{A}{v_{t/4} - A}$$

Thus the smaller V_t, the smaller the molecular weight and the greater the relative polarity.

If we define

$$\frac{r^3}{(r-d)^3} - 1 \equiv \pi$$

then if $p = \pi$, the protein is a spherical globule. Usually, $p \neq \pi$ and we can recognize two cases:

1. $p \gg \pi$ means a highly elongated protein, in other words, a fibrous protein.

2. $p < \pi$ means the p-residues do not cover the n-residues. This leaves n-residues on the surface and hydrophobic interaction may produce aggregation or clumping.

The above argument is supported by X-ray studies of myoglobin and hemoglobin showing that the 33 amino acid residues that are turned inside are nonpolar; on the other hand, the surface will show p-residues, plus those n-residues that have not "found a place" inside.

Additional insight into the protein folding problem comes from the work of Chothia, who has shown that for relatively small proteins ($M < 40 \times 10^3$ Daltons), the folding leads to reasonably well-defined relations between the area of the protein that is excluded from contact with water and the molecular weight of the protein. Let us define the accessible surface area (ASA) as the area over which contact between the atoms of the protein and the water molecules could occur. Strictly speaking, this term (ASA), originally introduced by Lee and Richards, means the area over which a water molecule can move without rupturing its van der Waals bond with some particular atom on the protein (and without penetrating the protein). If A_S is the ASA of the native protein and A_T is the ASA it would have if it were completely unfolded, then the ASA lost due to the folding of the protein is just

$$A_T - A_S \equiv A_B$$

and the proportion of the ASA buried by folding is

$$F_{AS} = A_B/A_T$$

Using data derived from X-ray diffraction studies of proteins, Chothia showed that

$$A_T = 1.44 M$$
$$F_{AS} = 0.595 + 0.536 M \times 10^{-5}$$

whence

$$A_B = 0.859 M + 0.774 M^2 \times 10^{-5}$$

From this, it is possible to estimate the hydrophobic contribution to the free energy of the folding process. Assuming an approximate value of 24 cal/A^2, Chothia found

$$F = 21 M + 19 M^2 \times 10^{-5}$$

This demonstration that the ASA of such proteins is determined by M is important because it establishes a common property for protein structure. Thus, although two small globular proteins may be completely different in function, detailed composition, and properties, if they have the same molecular weight, they will have the same ASA, as, for example, is found for α-chymotrypsin and concanavalin A. Chothia's work is also important because it serves to emphasize that the existence of the natural structure of these globular proteins requires the presence of water. In order to exhibit proper biological activity, the globular proteins must be at least slightly soluble. If the globular protein could form only secondary structure, the result would be that the polar groups would be generally removed from the surface. This would tend to produce a hydrophobic surface and thus a low-solubility structure. However, the protein also folds in such a way as to bury some of this hydrophobic surface, thereby increasing the solubility. The bigger the molecule, the higher the molecular weight and hence the more area that has to be removed from contact with water. In effect, globular proteins have evolved to have the least possible ASA for M, thus maximizing the free energy gained by folding.

Although Chothia's work is a major contribution to the understanding of native protein configuration, it should be pointed out that many difficulties remain to be removed before the behavior of globular proteins, as well as other macromolecules, can be understood. For example, it has been clear for some time that various experimental results can lead to rather contrasting pictures of the behavior of native macromolecules. Thus X-ray data strongly suggests that globular proteins are relatively tightly packed structures, and from this one would infer that the structure is therefore very stable and rather static. Since the thermal denaturation of such proteins occurs at a reasonably well-defined temperature, the tendency to see the X-ray data as suggesting a kind of "protein solid" is strongly reinforced. However, other experiments, chiefly using NMR, strongly suggest a more dynamic picture characterized by conformational changes. A large step to resolving this difficulty has been taken by Cooper, who has shown that in fact the internal energy of the molecule suffers large fluctuations, of the order of 6×10^{-20} cal/molecule. Obviously these fluctuations will not occur inphase through a protein solution. Therefore, the average fluctuation over the whole solution is essentially zero. Thus, experiments with low time resolution like X-ray diffraction emphasize the absence of dynamic behavior, while experimental techniques like NMR, capable of resolving rapidly varying properties, emphasize the dynamic behavior.

The shape of a protein under normal conditions appears to be fixed by the order of the amino acids as specified in protein synthesis process; the order of the amino acids is determined by the specific instructions coded in the cellular DNA. There are two basic views on this matter:

1. The protein folding represents an energy minimization process, and the final structure is that which produces an allover minimum energy for the protein.

2. The folding proceeds one step at a time; each step is determined only by the immediately favorable kinetics for that step. The "finished" protein is simply the resultant sum of the individual steps.

The viewpoint expressed in (1) is sometimes called the thermodynamic view, while that of (2) is known as the kinetic view. Recent work by Levitt seems to support the thermodynamic approach.

In Levitt's studies, the energy of the protein is described by the following function:

$$E = \sum_{\text{bonds}} \tfrac{1}{2} K_b (b_i - b^0)^2 + \sum_{\substack{\text{bond} \\ \text{angles}}} \tfrac{1}{2} k_\tau (\theta_i - \tau^0)^2$$

$$+ \sum_{\substack{\text{torsion} \\ \text{angles}}} k_\theta [1 - \cos(m\theta_i + \delta)]$$

$$+ \sum_{\substack{\text{nonbonded} \\ \text{distances} < R_{\max}}} [(A/r_i^{12} + B/r_i^{6})]$$

where

K_b = bond stretching force constant
k_τ = bond angle bending force constant
b^0 = bond lengths at equilibrium
τ^0 = bond angles at equilibrium
k_θ = energy required to twist bonds
m and δ are the period and phase, respectively
A and B are the repulsive and long-range force constants
r = atomic separation
θ = bond angle

The first two force constants are derived from the vibrational spectra of small molecules. Methods for evaluating the various parameters are given by Levitt. In Levitt's approach, computer iterations of the energy function are carried out until the function is minimized. In addition to eliminating stereochemical difficulties in models, this method can produce a set of coordinates in better agreement with the X-ray data than can be obtained from models.

The above approach appears to be able to predict the correct folding of globular proteins, as illustrated in Figure 3.3. Similar results have been obtained by Pipas, who has shown that the RNA cloverleaf is the structure with the lowest free energy of formation. An additional argument for the thermodynamic view comes from the X-ray structure analysis of the protein carboxypeptidase, an enzyme, in which the structure shows clearly that it is topologically impossible for the protein to have folded up from the end as it was being synthesized.

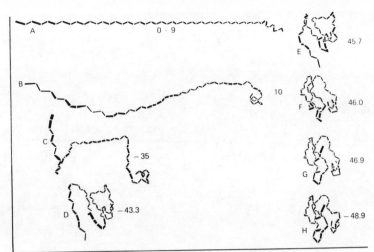

Figure 3.3 The above sketch is a summary of a computer simulation of a protein folding, according to Levitt and Washel. The first step is the extended chain seen at the top; the last is the conformation at the bottom. The energies are given in kcal/mol; note the steady decline toward the most stable state. The computational program used no information about the actual, final structure of the protein.

chap. 4

Basic En-
zyme Behav-
ior

The significance of proteins in living matter is already clear. Perhaps the most critical role of proteins is when they function as catalysts for the reactions necessary to maintain the living state; in this role, they are known as enzymes. Krebs, and those who have followed him, have now essentially proved that all chemical reactions in living cells are accompanied by enzyme catalysts. Why are the enzymes ubiquitous? Because the cell requires that essential organic macromolecules must be produced under conditions that no rightminded chemist would deliberately choose. Thus the synthetic reactions under these conditions would have vanishingly small rates were it not for the presence of the catalysts. From this, it is obvious that we cannot understand the chemical basis of life without understanding how the enzymes function.

PROPERTIES OF ENZYMES

The properties of those proteins that can act as enzymes may be summarized as follows:

1. They are globular, probably because this gives them stability in polar environments.
2. The catalytic process involves two distinct steps:
   (a) The binding reaction in which the substrate, which is the principal reactant, is favorably oriented.
   (b) The stereochemical reaction, occurring at the so-called active site of the enzyme. It is the stereochemical reaction that results in bond changes in the substrate.
3. The enzyme may undergo conformational changes, either before or after the stereochemical reaction.
4. Enzymes are, for uncertain reasons, high molecular weight proteins.
5. X-ray structure analysis shows that the active site is associated with a cleft in the structure of the protein. Furthermore, this active site is not necessarily formed by a portion of some particular strand of amino acids; contacts between various different strands of the protein may occur to form the active region.
6. Different enzymes with totally distinct structures may catalize the same reaction with the same substrate; however, the affinity between enzyme and substrate is always high.

DESCRIPTION OF ENZYME BEHAVIOR

The pioneer success in understanding the effect of an enzyme on a reaction rate came in 1913, with Michaelis and Menton's treatment of the enzyme invertase. Their work provided a model that is still the standard way of approaching the enzyme problem. They assumed the following reaction scheme:

$$\underbrace{E + S}_{\text{I}} \underset{k_2}{\overset{k_1}{\rightleftharpoons}} \underbrace{ES}_{\text{II}} \overset{k_0}{\to} \underbrace{E + P}_{\text{III}} \tag{4.1}$$

for the particular case of

$$\text{Sucrose} + H_2O \xrightarrow[\text{invertase}]{} \text{fructose} + \text{glucose}$$

Let C_E, C_S, C_{ES}, and C_P be the concentration in moles per liter of, respectively, the enzyme, the substrate, the enzyme-substrate complex, and the product. It is assumed that

$$C_S > C_E \qquad \text{and} \qquad C_P > C_E$$

So the formation of ES means that the concentration of the substrate is changed little. Therefore

$$C_S \simeq \text{constant}$$

and the concentration of free enzyme becomes

$$\simeq C_E - C_{ES}$$

Therefore, since

$$dP/dt = v = k_0 C_{ES} \simeq \text{constant}$$

and

$$\frac{dC_{ES}}{dt} = 0$$

(product is being formed, but never destroyed),

$$k_1 (C_E - C_{ES}) C_S = (k_2 + k_0) C_{ES}$$

A new constant, commonly known as the Michaelis constant, may now be defined:

$$K_m \equiv \frac{k_2 + k_0}{k_i}$$

so

$$K_m = \frac{(C_E - C_{ES})C_S}{C_{ES}}$$

whence

$$C_{ES} = \frac{C_E C_S}{K_m + C_S}$$

and therefore we have

$$\frac{dP}{dt} = v = k_0 C_{ES} = k_0 \frac{C_E C_S}{K_m + C_S}$$

which is the Michaelis-Menton equation. If the concentration of substrate is high,

$$C_S \gg K_m$$

and, consequently,

$$\frac{dP}{dt} \doteq v_{max} = k_0 C_E$$

which is the maximum rate of product formation. If the concentration of substrate is low,

$$\frac{dP}{dt} = \frac{k_0 C_E C_S}{K_m} = \frac{k_1 k_0 C_E C_S}{k_2 + k_0}$$

Suppose we consider the case when

$$dP/dt = \tfrac{1}{2}(dP/dt)_{max}$$

that is,

$$v = \tfrac{1}{2} v_{\max}$$

Then

$$\tfrac{1}{2}(dP/dt)_{\max} = \frac{(dP/dt)_{\max} C_S}{K_m + C_S}$$

so

$$\tfrac{1}{2} = C_S/CK_m + C_S)$$

and

$$K_m = C_S$$

Thus K_m is the substrate concentration at which dP/dt is half its maximum rate; this is independent of enzyme concentration. However, K_m is not a constant for a particular enzyme and can depend on substrate, pH, and T. Furthermore, if the enzyme can interact with different substrates, each will have a particular value of K_m. Thus the scheme of Equation 4.1 allows us to derive information on overall kinetics and maximum rates.

What is happening in the above reactions? If we consider a reaction in which two molecules approach each other, there is, generally, a repulsive force; therefore, there is a potential energy barrier. If ΔF is the free-energy difference per mole between the products and the reactants, then

$$K = [\text{products}]/[\text{reactants}] \ e^{-\Delta F/RT} \tag{4.1a}$$

Therefore, the reaction rate is proportional to the barrier height. Since enzymes accelerate the reaction rate, they must be "attacking" the barrier and either lowering it or providing an alternative path.

The scheme summarized in Equation 4.1 is actually too simple. A more realistic scheme would be

$$E + S \underset{k_{21}}{\overset{k_{12}}{\rightleftharpoons}} ES \underset{k_{32}}{\overset{k_{23}}{\rightleftharpoons}} EP \underset{k_{43}}{\overset{k_{34}}{\rightleftharpoons}} E + P \tag{4.2}$$

A second difficulty is the highly restrictive nature of the assumptions in the previous treatment; the concentration of the enzyme is often substantial compared to the concentration of the substrate.

On the assumption that a steady-state view of Equation 4.2 is valid, so that only a minimum number of steps is required, a treatment similar to that employed for Equation 4.1 may be carried out. The results are

$$V_{max}^P = \frac{k_{32}k_{21}}{k_{23} + k_{32} + k_{21}} C_E$$

and

$$K_m^P = K_m^S \frac{C_P V_{max}^P}{C_S V_{max}^S}$$

where

$$V_{max}^S = \frac{k_{23}k_{34}}{k_{23} + k_{32} + k_{34}} C_E$$

and

$$K_m^S = \frac{k_{21}(k_{32} + k_{34}) + k_{23}k_{34}}{k_{12}(k_{23} + k_{32} + k_{34})}$$

THE MECHANISM OF ENZYME BEHAVIOR

Can we get any physical insight into the enzyme process? How should we measure the impact of the enzyme? This is not as simple as one might think. For example, the turnover number may not be a very useful clue. Consider the reaction:

$$H_2O_2 + catalase \rightarrow H_2O + catalase + O_2$$

On the basis of turnover, catalase becomes an extremely potent enzyme. Yet, in fact, the reaction:

$$H_2O_2 \rightarrow H_2O + O_2$$

goes without any catalyst and, in addition, can be catalized by dust, glass beads, and several other very unimpressive materials. Do not be misled by this counterexample; the action of enzymes is both very selective and very impressive. For example, in the above reaction, if no catalyst is present, $\Delta F \approx 18 \times 10^3$ cal/mole. If platinum is used, the relative rate increases by 10^4, and ΔF becomes $\approx 12 \times 10^3$. The use of catalase, on the other hand, produces a relative rate increase of 10^7, with $\Delta F \approx 2 \times 10^3$. As another example, chitin, the material of the insect exoskeleton, is not at all hydrolized at 70° F by a solution of acetic and propionic acid. Yet, the addition to the solution of lysozyme, an enzyme in which the above acids are attached to its active site, produces rapid hydrolysis.

Koshland has argued that the most significant quantity for describing enzyme activity is

$$\frac{\text{Velocity with enzyme}}{\text{Velocity without enzyme}} \equiv v_e/v_0$$

and has given a simple model that suggests how the great rate increase may occur.

Consider an aqueous reaction:

$$A + B \rightleftharpoons AB \rightleftharpoons C + D$$

Let us assume the aqueous medium can be represented as a set of lattices. The reaction in the above equation requires a collision between A and B in order to form AB pairs. The molar concentration of AB is given by

$$\left\{ \frac{[A]\,[B]}{\text{Molar concentration of } H_2O} \right\}$$

multiplied by the coordination number, γ, or

$$\frac{[A]\,[B]}{55.5} \times \gamma$$

If no enzyme is present, the reaction rate is

$$V_0 = k_0\,[A]\,[B]$$

However, if an enzyme is present, the rate of the reaction will be determined by the enzyme concentration:

$$V_e = k_E [E]$$

Koshland now introduces a model in which the functional groups of the enzyme are characterized by two quantities: f, the fraction of the total number of molecules in nearest-neighbor sites and Θ, the angle over which correct orientation can be maintained between groups. Then, if $f \approx 1/\gamma$, $\Theta \approx 10°$, and the only contribution of the enzyme is to align the molecules so that the above is correct, Koshland shows that for the case of two substrates and one catalytic group,

$$\frac{v_e}{v_0} \approx 10^{10} \qquad \text{where } [A] = [B] \doteq 10^{-3} M$$
$$\text{and } [E] \doteq 10^{-5} M$$

In other words, a substantial increase in the reaction rate can be produced if the enzyme acts to correctly orient the reactant(s).

Since the enzyme is a protein, it is also a polyelectrolyte; that is, it exists in a number of charged states in solution. The number of different states is a function of the pH of the solution. In order to get a feel for this situation, consider a system:

$$AH_2 \underset{k_1}{\rightleftharpoons} AH^- \underset{k_2}{\rightleftharpoons} A^=$$

The total concentration is obviously the sum of the concentrations of the above species, since

$$k_1 = \frac{[AH^-][H^+]}{[AH_2]}$$

$$k_2 = \frac{[A^=][H^+]}{[AH^-]}$$

$$k_3 = \frac{[A^=][H_+]^2}{[AH_2] k_1}$$

We have

$$\Sigma \text{ all species} = \mathscr{A} = [AH_2] f = [AH_2^-] f^- = [A^=] f^=$$

where the f's are the so-called Michaelis functions:

$$f = 1 + k_1[H^+] + k_1 k_2/[H^+]^2$$

$$f^- = 1 + [H^+]/k_1 + k_2/[H^+]$$

$$f^= = 1 + [H^+]/k_2 + [H^+]^2/k_1 k_2$$

Let us assume that the active site of the enzyme functions in only one of the three possible states, say AH^-. Then

$$v = k[AH^-]$$

so

$$= \frac{k\mathscr{A}}{1 + [H^+]/k_1 + k_2/[H^+]}$$

$$= k\mathscr{A}\left[1 + \frac{[H]}{k_1} + \frac{k_2}{[H^+]} \right]^{-1}$$

We want to find the maximum, so differentiate the expression, set it equal to zero, and solve. We obtain

$$\partial v/\partial[H] = k\mathscr{A}\left(1 + \frac{[H]}{k_1} + \frac{k_2}{[H^+]} \right)^{-1} \left(\frac{1}{k_1} - H^{-2}k_2 \right) = 0$$

$$\therefore \frac{k\mathscr{A}}{1 + [H]/k_1 + k_2/[H]} \cdot \left\{ \frac{1}{k_1} - \frac{k_2}{[H]^2} \right\} = 0$$

whence

$$\frac{1}{k_1} = \frac{k_2}{[H]^2}$$

and

$$[H] = \sqrt{k_1 k_2}$$

thereby giving, for this simple model, the dependence of enzyme activity on pH, since

$$pH = \log[H]$$

The polyelectrolyte character has a second consequence. The free energies of the different ionized states are essentially the same and, therefore, charge motion in the complexes is not difficult. Such charge motion could lead to electrostatic interaction between the enzyme and substrate. It has been suggested, originally by Kirkwood, that such a gain in energy is the source of the decrease in the barrier height referred to in Equation 4.1a.

Consider an enzyme-substrate complex with a potential energy u. Then

$$e^{-\overline{u/kT}} = e^{-w/kT}$$

and the average of all the fluctuating states should be the energy of the system. Thus

$$w = \langle u \rangle - \tfrac{1}{2}kT \left[\langle u^2 \rangle - \langle u \rangle^2 \right] + 2^d \text{ order contributions}$$

If no fluctuations occur, $\langle u^2 \rangle \equiv \langle u \rangle^2$ and $w = \langle u \rangle$.

Suppose we take as a model a protein with n basic groups; the charge on each group is $z_n e$. The distance between a group of the active site and the substrate is R_i. Then

$$u = \frac{\Sigma z_i \, e(\text{dipole moment of substrate}) \cos \measuredangle (R, \text{dipole})}{\epsilon \, R_i^2}$$

where ϵ is the dielectric constant. It can be shown that

$$\omega \simeq - \frac{n_\alpha e^2 p^2}{4\epsilon^2 r_\alpha^4 kT} \frac{K[H^+]}{([H^+] + K)^2}$$

where n_α is the number of adjacent basic groups. Kirkwood suggested a value of the local dielectric constant of $\epsilon \simeq 10$. Since the dielectric constant for water is $\epsilon \doteq 80$, such a large difference needs a quantitative defense, which is very hard to produce. Although the above does lead to a Gaussian dependence of k on pH, there is really no reason to consider that conclusive. Furthermore, other predictions of this model, such as particular relaxation components in the spectrum, are not found.

No theory can yet predict the dependence of the enzyme structure on pH. However, the effect of pH on the secondary structure is to some degree understandable. For example, Zimm and Rice have shown that in the case of helical polyaminoacids composed of ionizable groups, the transformation between helix and random coil, which occurs at pH $\simeq$ pK of the group, comes about through the production of charges on the units of the chain. This gives rise to electrostatic

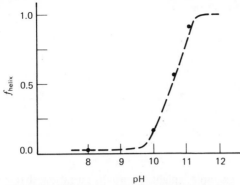

Figure 4.1 This graph shows the sharpness of the transition between a helix and a random coil with change in pH.

repulsion between the chains and this force unfolds the helix. Thus we have the sharp dependence of the fractional helicity on pH shown in Figure 4.1.

A major source of difficulty in the analysis of enzyme action is that studies carried out under equilibrium conditions are almost certainly incapable of providing enough information to permit us to discriminate between various possible mechanisms. Theoretical mechanisms are really tested only by a demand for kinetic details, but we are hampered by our ignorance of the exact details regarding the intermediate states. The most important contribution to the solution of this problem has been made by Eigen's studies of relaxation spectra. In these studies, a system is perturbed by subjecting it to a very sudden excursion in temperature or pressure. By keeping the T- or P-jumps small, the behavior of the system can be approximated by a set of linear differential equations whose solutions contain characteristic time constants. These time constants are the relaxation times of the system, and the observed quantity is a sum of exponentials in these constants. The deduced set of constants is the spectrum of the relaxation times.

For example, suppose we have a set of rate equations. These may be linearized so that if we are observing a variable $\phi(t)$, we may write

$$\phi(t) = \sum_i \beta_i e^{-t/\tau_i}$$

where τ_i is by definition the relaxation time, and the expression has been normalized by making

$$\sum_i \beta_i = 1$$

Then the average relaxation time is

$$\bar{\tau} = \sum \beta_i \tau_i$$

which is measurable as

$$\bar{\tau} = \int_0^\infty \phi(t)dt$$

The function ϕ, for example, might be a concentration that would be monitored spectroscopically by measuring the strength of an absorption feature in the solution in which the reaction was occurring.

The method for observing $\phi(t)$ in an enzyme system is by means of a single "jolt" of energy to produce a very quick temperature increase (the T-jump), followed by a relaxation to equilibrium. The most efficient means for producing the T-jump is by a capacitative discharge across a spark gap in the observation cell or by the heating produced by short bursts of microwaves directed at the cell. In either case, heating is very rapid; for example, the microwave method produces a T-jump in about half a microsecond. The apparatus, along with some results, is shown in Figure 4.2.

By providing detailed structure, some of which are shown in Figure 4.3, X-ray diffraction studies have also played an important role in enzyme studies. This has been possible because enzymes are globular proteins, which are not too difficult to crystallize, although if allowed to dry they will become disordered.

The X-ray structural results generally confirm the "lock and key" model of enzyme-substrate union, originally proposed by Fisher, and illustrated in Figure 4.4. The X-ray results also clearly show how van der Waals forces and hydrogen bonding can bind the substrate to the enzyme in the active site. This process is made more understandable by the fact that the X-ray results, admittedly only for a relatively few

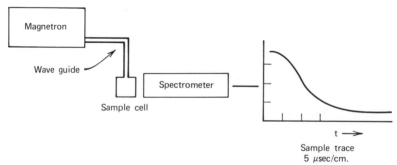

Figure 4.2 This drawing shows a schematic arrangement of the apparatus for T-jump experiments.

enzymes, show a common feature for all, namely that the polar residues are essentially excluded from the interior of the enzyme. Thus the interior is effectively nonpolar, the possible electrical forces unreduced by the presence of water, and the activity of the enzyme presumably susceptible to analysis by nothing more complicated than electrostatics.

Fisher's "lock and key" hypothesis has been confirmed by X-ray diffraction in the sense that the substrate is correctly fitted at the active site. However, this model does not require that the shape of the active site is identical in the bound and unbound states. Koshland has developed a model of a flexible enzyme in which the fit is induced by the binding process of the substrate or by the combined binding of the substrate and another small molecule. Proof that enzyme conformations do change upon the binding of small molecules has been obtained by X-ray diffraction. This model explains the apparent exceptions to the "lock and key," such as compounds that bind to the enzyme without any reaction occurring. This would now be seen as a case in which the binding changed the conformation of the enzyme in such a way that the active groups could not correctly align, thereby preventing any reaction from taking place.

Furthermore, new techniques promise to extend our knowledge of the enzymatic reaction. One such new approach is the study of enzyme reactions at low temperatures. In 1970, Douzou showed that enzymes still functioned in a biologically significant way at low temperatures and that perfectly proper enzyme-substrate intermediates could be isolated in stable form at low temperatures, provided that the correct solvent systems were utilized. One obvious advantage of this approach is that the rapidity of the reaction is greatly reduced. Another less obvious advantage, pointed out by Fink, is that by using different low temperatures, various intermediates in the reaction can be trapped and isolated. Thus, if a reaction is begun at some low temperature and then the temperature is gradually raised, the total enzyme-substrate reaction will appear as a set of reaction steps. Also, the various intermediate steps are thus stabilized, and the complexes characterizing those states can then be subjected to X-ray diffraction analysis to determine their structure.

The understanding of *in vitro* enzyme reactions is therefore a situation in which many important aspects are no longer uncertain: we know several enzyme structures that include the active site, we know the intermediates in many important enzyme-substrate reactions, and we know quite a bit about the energetics of the process. These very real accomplishments should not, however, lead to the conclusion that the enzyme problem is effectively solved. Instead, Wooley's comment on carbonic anhydrase provides an appropriate perspective:

> Detected in 1928, discovered in 1932, . . . obtained pure in 1959, crystallized in 1962 and the subject of some 1800 papers and articles, carbonic anhydrase continues to elude a full mechanistic description, even though this enzyme is small, stable, easily isolated and its X-ray structure is known to a resolution of 2Å.

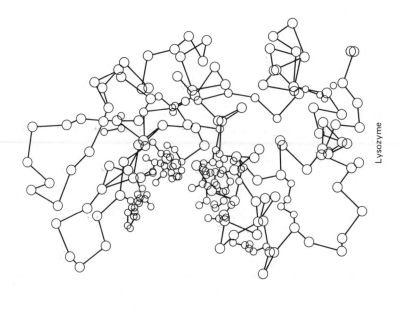

Lysozyme

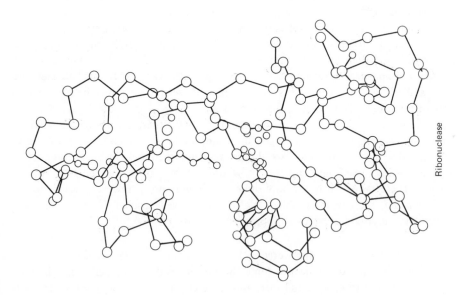

Ribonuclease

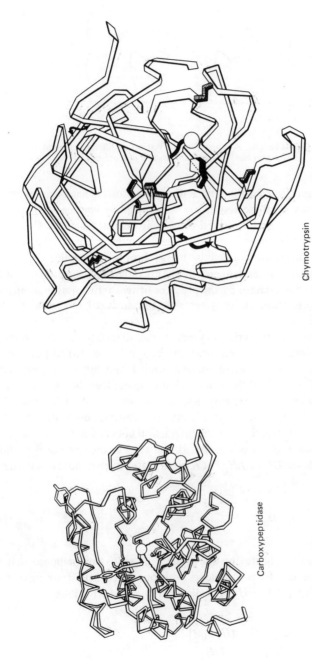

Chymotrypsin

Carboxypeptidase

Figure 4.3 Shown above are some enzyme structures determined by X-ray analysis.

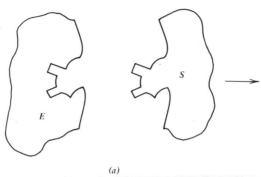

(a)

Figure 4.4a A schematic illustration of Fisher's original "lock and key" mechanism for enzyme action.

Although much of what we have discussed thus far has been connected with *in vitro* studies, many of these results are valid in an *in vivo* context, and the kinetic schemes provide an accurate description of many cellular reactions. Of course, some *in vivo* processes may be more complicated and require a more detailed analysis. For example, many components of cellular metabolism are controlled through a variation in the catalytic activity of certain enzymes. This variation is the property of a particular class of enzymes characterized by the presence of two active sites that can interact. The properties of these allosteric enzymes was first analyzed quantitatively by Monod, Wyman, and Changeux.

The activity of allosteric enzymes is not described by the simple Michaelis-Menten scheme, and allosteric enzymes can be quickly identified because the plot of reaction velocity versus substrate concentration is not hyperbolic, as predicted by a Michaelis-Menten analysis of the kinetics, but is sigmoid, as shown in Figure 4.5.

In the analysis of these enzymes, it is assumed that each active site has two states, H and L; state H has a high affinity for substrate, but state L has very little or no affinity. Furthermore, through the interaction between the active sites, the enzyme is either in one state or the other; that is, the state of the enzyme as a whole is either HH or LL, but never HL or LH. Let R be the ratio of enzyme states in the absence of substrate:

$$R \equiv L_0/H_0$$

where the subscript indicates no substrate is present. Now suppose that in the HH state, the enzyme can form complexes with one or two substrate molecules, so that $H_0 + M \rightarrow H_1$ and $H_1 + M \rightarrow H_2$. The dissociation constant is then

$$k = \frac{2[H_0][M]}{[H_1]} = \frac{[H_1][M]}{2[H_2]}$$

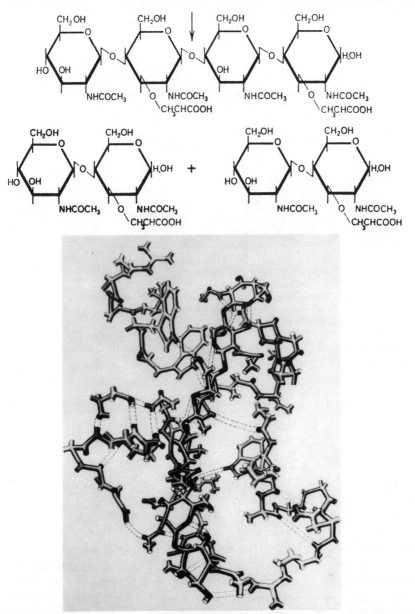

Figure 4.4b The active site of the enzyme lysozyme, as determined by X-ray diffraction is shown. This enzyme interacts with the substrate tetrasaccharide, shown above the enzyme. The action of the enzyme is to split the substrate into two parts by cleaving the oxygen bond indicated with the arrow. This is the basis for lysozymes' role in the phagocytes, mobile cells that circulate in the body and destroy invading bacteria. Since the tetrasaccharide is an integral part of the bacterial cell wall, splitting the molecule causes the cell wall to fall apart.

95

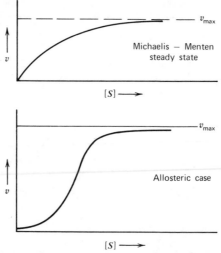

Figure 4.5 These two graphs show how the shape of the curve obtained by plotting reaction rate versus substrate concentration changes when a conventional Michaelis–Menton scheme is replaced by allosteric behavior.

The fractional saturation of the enzyme population is then the ratio of the number of occupied states to the total number of states:

$$F = \frac{[\text{occupied states}]}{[\text{total states}]} = \frac{[H_1] + 2[H_2]}{2([L_0] + [H_0] + [H_1] + [H_2])}$$

where

$$F = \frac{[M]/k(1 + [M]/k)}{R + (1 + [M]/k)^2}$$

Typical values of k are $\sim 10^{-5}$ M; $R \approx 10^4$. If we now plot F versus $[M]$, a sigmoid curve is obtained. The great success of this model is that it opens the way to understanding how enzyme rates may be controlled through the binding to the enzyme of an inhibitory or an activating substance. In the first case, the binding is preferential to L; in the second, to H. Thus inhibitors shift the ratio of states toward L and activators, toward H. Such effects clearly can be analyzed as a shift in the equilibrium constant of the reaction and appear to the observer as variations in the catalytic activity of the enzyme. In the cell, an enzyme reaction can be controlled by having either the intermediate or the product itself function as inhibitors while the substrate serves as an activator.

The allosteric control model, and essentially all of the quantitative work based on it, assumes that the control mechanism is centered on the interaction between a single enzyme molecule and a smaller substrate molecule. Nevertheless, recent work by Cohen and Benedek has shown that there is at least one important case in which reversible polymerization of the enzyme appears to be the control mechanism. They considered the case of glutamate dehydrogenase (GDH). This enzyme is found in the mitochondria and is composed of six identical subunits. It is involved in a key reaction in which proteins are catabolized to provide cellular energy. The enzyme is inhibited by two energy supplying molecules, ATP and GTP (guanosine triphosphate), but activated by two reduced forms, ADP (adenosine diphosphate) and GDP (guanosine diphosphate). Hence, when the cellular energy storage is "up," there is a lot of ATP or GTP around and the reaction slows. However, if the cellular energy is low, ATP and GTP have been reduced to ADP and GDP, and the reaction accelerates, increasing the oxidation of amino acids and supplying energy to the cell.

The GDH system is a warning not to make the models of enzyme behavior too restrictive. Although many in vitro experiments are carried out at low concentrations and although some of the kinetic schemes discussed before assumed such concentrations, there are enzymes that function at high concentration in vivo. GDH is one such enzyme and occurs in the cell at concentrations on the order of milligrams per milliliters.

In Cohen and Benedek's model, GDH is assumed to have two forms, one active and one inactive; furthermore, the active form is more susceptible to polymerization. It is significant for this view that GDH will polymerize in vitro at concentrations on the order of milligrams per milliliters and that this polymerization is reversible by lowering the concentration; also the addition of ATP reduces polymerization while ADP increases it. This polymerization is important in the cellular context because it changes the ratio of the number of active to the number of inactive molecules. This leads to an overall enzyme activity that is very sharply dependent on the concentration of the enzyme when that concentration is high, but not very dependent when that concentration is low.

By using laser light scattering measurements, Cohen and Benedek show that linear polymerization of the enzyme does indeed occur, and that the polymers are a mix of the active and inactive forms. Thus the oxidation rate of protein is regulated by the increasing or decreasing polymerization of the GDH enzyme, which is itself responding to the relative abundance or scarcity of energy supplying molecules.

chap. 5

The Elementary Properties Of Nucleic Acid

The nucleic acids, as a class of biologically significant macromolecules, were recognized long before their specific association with the genetic process became clear. The discovery of that role began in 1928 with Griffith's demonstration that nonpathogenic pneumococcus bacteria could be transformed into pathogenic bacteria through treatment with a cell-free extract obtained from the pathogens. This transformation occurred when the nonpathogens became able to synthesize a polysaccharide cell coating. In 1944 Avery, MacLeod, and McCarty showed that the active component of the cell-free extract, the so-called transforming principle, was, in fact, DNA. This was a crucial result because, until this work was done, protein had been assumed to be the genetic messenger.

X-RAY STRUCTURE

Although the great significance of the above results is now, in hindsight, very clear, at the time, the impact was more modest and the growth of interest in nucleic acid structures was relatively slow. However, that interest, centered on DNA, eventually led to two important results. One was Wilkins' and Franklin's successful efforts at obtaining good X-ray diffraction pictures, and the other was Chargaff's discovery, in 1950, that the ratio of the bases adenine (A) to thymine (T) and guanine (G) to cytosine (C) were both unity, regardless of the source of the DNA.

Guided by Chargaff's result, Crick and Watson used models to predict the X-ray patterns; the correct model of DNA was then the one that satisfied the various chemical requirements and predicted a diffraction pattern that matched the actual X-ray diffraction pattern of ordered DNA. This approach was necessary because of the

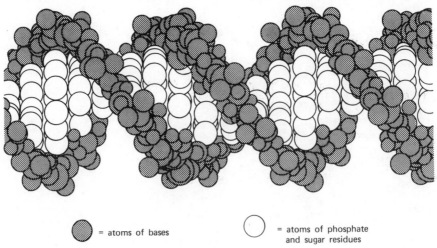

 = atoms of bases   = atoms of phosphate and sugar residues

Figure 5.1 The DNA double-strand model.

structural complexity of the macromolecule and because of the fact that, like fibrous proteins, DNA does not crystallize very well.

As is well known, the correct model for DNA is found to be a double helix. The backbone of the macromolecule is formed from deoxyriboses, which are connected by phosphodiester bridges (Figure 5.1). This structure is common to all DNA macromolecules. The various kinds of DNA are differentiated by the sequence of the bases. How does this general structure give rise to the observed X-ray diffraction pattern of DNA? The pattern, shown in Figure 5.2, is produced by the turns of the helix, which

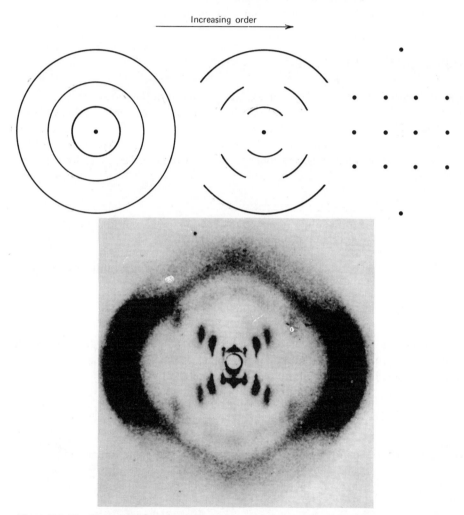

Figure 5.2 The X-ray diffraction pattern changes markedly as the order of the crystal increases, as shown in the drawing. The X-ray diffraction pattern for DNA should be understood in that light.

act like the slits of two gratings inclined to each other. The net effect is an x-shaped array of smears, rather than a precise arrangement of sharply defined spots. From the angle between the arms of the "x" and the spacing of the smears, the pitch and radius of the helix may be deduced. One of the "nice" features of DNA is that phosphorus occurs in the alternating phosphate and sugar groups from which the helix is formed. Phosphorus is a reasonably "heavy" atom and therefore a relatively good X-ray scatterer. Since the two strands that make up the helix are interwoven so that no translation motion of one with respect to the other can occur, because of the linking of the purine and pyrimidines, the phosphorus atoms produce a regular pattern in the array; thus DNA has, in effect, already got a substituted heavy atom.

ORIGIN OF THE STRUCTURE

Can the general structure of DNA be explained in terms of the forces operating in the macromolecule? To answer this question, and others about the conformation of DNA, Pullman and his co-workers have carried out extensive quantum mechanical calculations. In order to understand this work, it is necessary to recall that in 1926 Schrödinger argued that it was possible to define a function, ψ, generally complex, such that

$$\frac{\partial^2 \psi}{\partial x^2} + \frac{\partial^2 \psi}{\partial y^2} + \frac{\partial^2 \psi}{\partial z^2} = -\frac{4\pi^2}{\lambda^2} \psi$$

where λ is the deBroglie wavelength, connected to the momentum of the system by

$$\lambda = h/p$$

According to the interpretation developed by Bohr, the wave function ψ is a guide to the probability of finding various values of the physical quantities that we wish to measure in order to describe the system. Specifically,

$$\psi \psi^* d\tau$$

is proportional to the probability of localizing the system in some volume, and

$$\langle x \rangle = \int \psi \times \psi^* dx / \int \psi \psi^* dx$$

is the expectation value of some quantity x, for example, energy or angular momentum, associated with the system.

These ideas can be extended to calculate the properties both of atoms with many electrons and of molecules composed of many atoms and very large numbers of electrons. A method for doing this was originally developed by Mulliken; in this approach the Schrödinger equation, or a more convenient form of it, is solved for separate electrons. The result is a description of the electron in terms of its energy, orbital angular momentum, etc. This solution is known as an orbital and can be represented pictorially as a charge distribution; this distribution reflects the fact that a particular electron is more likely to be found in certain regions of the system than in others. The behavior of the electron cloud is then deduced by the proper combination of the orbitals. Thus, the wave function of a multielectron system can be obtained by the judicious combination of individual one-electron wave functions, carried out according to one of several possible mathematical procedures.

Pullman has applied these methods to the nucleic acid bases. One the basis of his results it is possible to give the distribution of charge on each base and to calculate the approximate dipole moments. These results are shown in Figure 5.3.

Now we need to find the interaction energy between the bases; the pairing of the bases is well known, and hence we are led to ask how the members of a pair can "recognize" each other. Can we find an arrangement of forces that is favorable to the formation of exactly the pairs that occur?

In Chapter 4, it was pointed out that one force that acted between molecules was due to the dipole–dipole interaction. As Figure 5.3 shows, the dipole moments of the pairs are significant. Furthermore, experimental evidence strongly suggests that the force between the base pairs owes more to the polar interaction than does the force between the bases stacked along the helix. This statement should not be over-interpreted. Any force can be resolved into components in different directions, and the notion that base pairs are due solely to one kind of force and base stacking to another will not survive criticism.

With this caveat, we now observe that the dipole interactions between the base

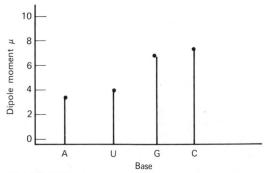

Figure 5.3 The bar graph indicates the relative dipole moments of the four bases in DNA, as calculated by quantum mechanics.

pairs can be calculated in two different ways. In one approach, the interaction energy is taken as the total contribution of three distinct sources:

$$E = \text{dipole–dipole interaction} + \text{dipole-induced dipole interaction}$$
$$+ \text{London force}$$

The last term in the above equation originates in the mutual perturbation of the electron clouds and is mainly an interaction between instantaneous dipole moments. If the molecules are in the ground state, the London force is always attractive. The first analysis of this force was carried out by the physicist Fritz London — hence the name. Denoting the dipole moments and polarizabilities as μ and α, the separation as R, and the ionization potential as I, the equation for the dipole interaction energy between two bases X and Y is then

$$E = \frac{\mu_X \cdot \mu_Y - 3(\mu_X \cdot \mathbf{R})(\mu_Y \cdot \mathbf{R})/R^2}{R^3}$$
$$+ \frac{\alpha_X [3(\mu_Y \cdot \mathbf{R}/R) + 1] + \alpha_Y [3(\mu_X \cdot \mathbf{R}/R) + 1]}{2R^6} + \frac{3I_X I_Y \alpha_X \alpha_Y}{2(I_X + I_Y)R^6}$$

A second way to proceed is to calculate the force between a pair of bases as due to the interaction between separate positive and negative charges, plus contributions from charge-induced dipoles and the London force. The expression for the London force has already been given as the third term in the last equation. The expressions for the charge interaction and the induced dipole contributions are

$$E_{\substack{\text{charge} \\ \text{interaction}}} = \sum_{\substack{\text{all} \\ \text{combinations}}} \frac{(\text{charges in } X) \times (\text{charges in } Y)}{\text{separation of charges}}$$

$$E_{\substack{\text{induced} \\ \text{dipole}}} = -\tfrac{1}{2}(\mu_{XY}E_{XY} + \mu_{YX}E_{YX})$$

$$\mathbf{E}_{XY} = \sum (\rho_X/R^3)\mathbf{R}$$

$$\mathbf{E}_{YX} = \sum (\rho_Y/R^3)\mathbf{R}$$

where μ_{XY} is the dipole induced in the base Y by the charges on the base X and $\mathbf{E}_{XY}$ is the electric field produced at Y (at the location of the induced dipole) by charges on X, with similar meanings for μ_{YX} and $\mathbf{E}_{YX}$.

Of the two methods for calculating the dipole interaction energy between bases, the second method gives more accurate results. For example, the results from the second method indicate energy minima at places where hydrogen bonding could in

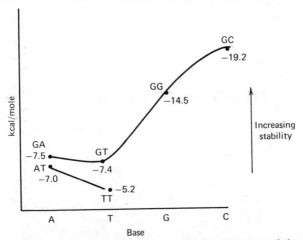

Figure 5.4 This graph shows the results of calculations of the binding energy between various DNA bases, as carried out by quantum mechanics. In the lower curve, the binding energy is shown for the formation of a base pair from A and T and from T and T. In the upper curve the energy is calculated for the case of G binding to each of the four bases. In each set, the most stable base pair is the one with the strongest binding (i.e., the most negative binding energy), and in each case that base pair is the one that actually occurs in double-strand DNA, that is, AT and GC.

fact occur. Using the second method, Pullman has calculated energies for various combinations of base pairs. The more negative the energy, the tighter the binding of the bases and the more favorable the interaction. Some of these results are shown in Figure 5.4; the agreement between the results and the observed favorable pairs in real DNA is very encouraging.

The second contribution to the interaction energy, that of the stacking order, is a vastly more difficult problem. However, it is clear that the total interaction energy among the bases has an important contribution from the stacking, the order in which the bases are arranged along the sequence.

The third contribution to the interaction energy occurs when the nucleic acid is in solution. In water, hydrogen bonding between the water molecules leads to the formation of short-lived, small-scale ordering. The effect is often described as a "flicker phenomena"; there is always some medium-scale structure present in the water, but the specific molecules that are involved in making up that structure and the specific sites that those molecules occupy is in a state of constant flux. Nevertheless, the nucleic acid base will "see" itself surrounded by a lattice that it locally disrupts. In order to form a sequence, that is, a stack of bases, a hole must be made in the water lattice. Since many unordered bases will require more disordering of the water than will a set of stacked bases, the entropy of the stacked arrangement is smaller than the entropy of the random individual bases. Therefore, the stacked sequence is a more energetically favorable situation. The above situation should not be confused with the more complicated possibility, of uncertain importance, that the water molecules immediately adjacent to the stacked bases actually bind to them.

SOME PROPERTIES OF DNA

Can we calculate the change in heat content for the formation of a base pair? Applequist has shown that

$$\Delta H = \frac{\ln \beta C}{\left(\dfrac{1}{T_m} - \dfrac{1}{T_m^{\infty}} \right)} \frac{1}{R}$$

where

T_m is the melting temperature for a known group

T_m^{∞} is the melting temperature for a chain of such groups

C is the concentration

β is the equilibrium constant for forming the first pair

Therefore, by measuring T_m as a function of concentration, H may be found. It has also been shown that the proportion of bases, f, that are paired depends on the temperature according to

$$\frac{df}{dt} = \frac{N \Delta H}{6 R T_m^2}$$

and from spectroscopic studies, it is known that

$$\Delta H \simeq 8 \text{ kcal/mole}$$

in order to form a base pair in the helix. Therefore, if there are N bases, the whole helix requires

$$-H = (N - 1) \, 8 \text{ kcal/mole}$$

where $N - 1$ occurs because the first base pair is not involved in any stacking.

Quantum mechanical methods can also be used to find how the total energy of the molecule depends on various arrangements of the atoms in the molecule. Of course, the arrangements studied are chosen to fulfill the various chemical requirements. Such energy calculations can yield the energetically most favored arrangement and thus predict the actual molecular conformation.

Not all of the techniques for carrying out these calculations are equally successful. If we limit ourselves to the best results, we find that they are consistent with the experimental data. This is particularly encouraging, since the computations deal with free molecules, while the experimental results come from studies of solutions of crystals. This agreement suggests that no extensive (and therefore probably very complicated) interactions occur between these macromolecules in biological situations.

This conclusion is supported by recent NMR studies on yeast tRNA carried out by Robillard and his colleagues. The choice of tRNA was deliberate. Although NMR has been used for some time in the study of proteins, NMR studies of nucleic acids are relatively recent. This is because the nucleic acids of interest are of high molecular weight but contain only four major components. Thus, the NMR spectrum is a very crowded collection of unresolvable, uninterpretable lines. Therefore, the most favorable case is the molecule with the lowest molecular weight. Robillard used the X-ray data for tRNA to calculate certain features of the NMR spectrum; the features are in fact observed. Since the X-ray data is derived from crystals while the NMR spectrum is obtained from solutions, and since the quantum mechanical calculations show that the conformations that are most likely for the free molecules (and are therefore the most stable), are the ones found in the crystal forms, we can begin to believe that the *in vitro* results are correct representations of the *in vivo* situation.

Actually, although the X-ray results are irreplaceable for the determination of the dominant structure, once the structure is known, the quantum mechanical results become potentially more valuable. The reason for this is that NMR clearly shows that in most cases a variety of conformations are present in solution. Crystallizing presumably allows the most favorable form to dominate. However, the quantum mechanical results yield energies of the various possible stable conformations and thus the energy required to pass from one form to another.

Such work is already enlarging our view of the dynamics of nucleic acids in solution. At one time, the nucleic acids were thought to be rather rigid. This was based on the argument that since mononucleotides were rigid, the polynucleotides, and hence the functional nucleic acids, would also be fairly stiff except under very special conditions. This did not seem unreasonable, since the phosphate group could be a strong contributor to the rigidity of the nucleotide. Furthermore, any functional motion could occur through rotation around the phosphodiester bonds. This view is now probably too simple. Recent work by Evans and Sarina has in fact shown that the nucleotides of uridine and cytosine are stiffer than the nucleosides.

These various lines of evidence are thus suggestive of nucleic acid flexibility, and this property is clearly important if alterations in conformations are found to be important in the cellular context, as now seems to be the case.

Much valuable data on the nucleic acids can be derived from ultraviolet (UV) spectra. One of the most useful results of the molecular orbital calculations is the prediction of the electronic structure and from that the UV spectral features. Pyrimidine has three strong bands at 267, 240, and 180 nm, which show many changes under small structural or chemical influences. The spectra of purines is much more complicated and sensitive to substitutions. In both cases, however, spectra can give answers to questions about the concentration in solution and the identification and determination of molecular structure.

A useful source of information about the primary structure of DNA comes from centrifugation studies. Schildkraut, et al., have shown that the GC content of DNA can be determined from the density by the relation:

$$GC = \frac{\rho - 1.660}{0.098} \times 10^2$$

The simplest technique is to include a nucleic acid of known density in the sample to be centrifuged. Then we have

$$\rho_{unknown}^{DNA} = \rho_{known}^{DNA} + \frac{\omega^2}{2} (r_s^2 - r_m^2)$$

where ω is the angular velocity and $r_s - r_m$ would be the difference in the distance from the axis of the sample and the known DNA. This result can be correlated with melting curve studies that show:

$$GC = (T_m - 69.3) \times 2.44 \qquad \text{(mole percentage)}$$

The Messelsohn-Stahl experiments show clearly that a separation of the native strands of the double helix is a necessary step in the dynamics of cellular DNA. How does this necessary denaturation occur? The fact that there must be a force acting can be seen by assuming that no force is present and that the unwinding is a purely random process. At T_m, one calculates that about 20 minutes per 10^4 pairs is required; the experimental value is in fact a few hundredths of a second.

Although the problem remains difficult and of uncertain solution, the most likely physical explanation is that the separation of a specific base pair is accompanied by a positive entropy change, and that this entropy change then drives the unwinding of the helix.

Melting curve studies can provide insight into the effects of DNA structure on the denaturation process. Recall that the helix is stabilized by hydrogen bonds between the A and T bases (~ -2 kcal) and between the G and C bases (~ -3 kcal), as well as by the stacking forces between the pyrimidine and purine rings (~ -7 kcal/pair).

Denaturation is thus a three-step process. First, there is a collapse of the hydrogen bonded double helix structure; then, there is a collapse of the base stacking order; and finally, there is strand separation. This process can be followed by optical spectroscopy at 260 nm. Physically, what one is observing is a cooperative helix coil transition with a sharp melting curve; the temperature range over which the transition actually occurs is $\Delta T \approx 10°C$.

Theory indicates that in order to see such a transition there must be at least 500 base pairs linking the two strands. Furthermore, only very small regions composed of just AT or GC pairs can be present. Obviously, if there were a long run of say AT pairs and then a long run of GC pairs, a melting curve with two sharp steps would be observed because the net bonding energies of the two regions would be different. However, DNA can have small regions with uniform base pairs. Therefore, one might expect "fine structure" in the melting curve. Wada and Tachibana have, in fact, been able to detect just such an effect, and the observed fine structure agrees with the composition of the various regions of the DNA as determined spectroscopically.

Additional information about nucleic acid processes can be gained from hybridization experiments developed by Spiegelman in 1960 in which a mixture of nucleic acid single strands are gently heated in order to stimulate the formation of double strands, which are then isolated. The hybrid yield is then a measure of the occurrence of regions with complementary base pairs in the original collection of single strands. This experiment can be analyzed as follows. Assume the reaction is

$$M + N \underset{k_2}{\overset{k_1}{\rightleftharpoons}} MN$$

where M and N are, respectively, the bases on each strand. Then a simple kinetic picture would be described by

$$\frac{dx}{dt} = k_1(m - x)(n - x) - k_2 x$$

where $m = [M]$ and $x = $ [double strands]. We start with

$$m = n$$

and also

$$k_1 \gg k_2$$

that is, once a helix is formed, it is very unlikely to unwind. Therefore,

$$dx/dt = k_1(m - x)^2$$

whence

$$x/m = mk_1 t/(1 + mk_1 t)$$

with a half life of

$$\tau = 1/k_1 m$$

Suppose $n \neq m$, say $n \ll m$. Then

$$dx/dt = k_1 n(m - x) - k_2 x$$

whence

$$k_1 b + k_2 = \frac{1}{t} \left(\ln r_n/r_n - x \right)$$

Where b is the constant of integration:

$$\frac{1}{k_1} \left(\frac{\ln 2}{\ln \dfrac{2}{k_1 m + k_2}} \right) - \frac{k_2}{k_1}$$

and r_n is the amount of N bound at equilibrium; now

$$\tau = \ln[2/(k_1 m + k_2)]$$

We may now use

$$k_1 \gg k_2$$

to get

$$t \doteq (1/k_1 m) \ln n/n - x$$

and

$$\tau \doteq \ln 2/k_1 m$$

One result of these hybridization studies confirms one of the main points in the "central dogma." According to that scheme, the transcription of the information coded in the DNA strand into a set of amino acids linked to form a protein requires the production of messenger RNA (mRNA). Thus, mRNA, which is single stranded, should have a base series complementary to the base series on the DNA template. This was shown in the following way. First mRNA, which should be complementary to the DNA in the bacteriophage $T2$, was isolated from a population of $E.$ $coli$, which had been infected with $T2$. This mRNA was labelled with P^{32}. Separately, DNA from $T2$ was isolated and labelled with tritium. These nucleic acids were then mixed and heated to ~100°C, which causes the denaturation of the DNA into separate strands. Upon cooling, the sample was centrifuged for about three days. Three components were found: At maximum density was single-strand mRNA; then came renatured, double-strand DNA; and finally, a mix of mRNA-DNA, that is, a hybrid combination of a single strand of DNA with a single strand of mRNA. However, if the experiment is now done for DNAs from other bacteria or virus particles, the mRNA does not form a hybrid with any of this DNA, even if the overall base ratios are similar to those in the mRNA. Obviously, hybridization requires not only the proper general base ratios but also exactly the proper base order and can only occur when the mRNA is complementary to the DNA template.

DNA IN A CELLULAR CONTEXT

Let us discuss the behavior of nucleic acid in a cellular context. A key observation with respect to bacterial nucleic acid was provided by the work of Schaecter et al., which demonstrated that (1) the average cell mass increases exponentially with the growth rate; (2) the average amount of RNA per cell also increases exponentially, but somewhat more rapidly than (1); and (3) the rate of increase of DNA per cell is somewhat less than (1). Therefore the significant parameter in cell size and composition, in the broadest sense, is the growth rate; the particular growth media by which that growth rate is maintained is relatively insignificant.

In the case of $E.$ $coli$, a variety of experiments indicate that nearly all the DNA is in the form of a ring of double helix. This is also the case for the other six bacterial species that have been examined. The DNA replicates from a specific point, often called the chromosome origin. The elongation of the chain goes in both directions around the ring from the origin, until the so-called "replication forks" meet halfway around the ring; this implies that the "fork velocity" is equal on each half of the ring. The time required for the fork to travel halfway around the ring is denoted by C.

There are then two questions:

1. How often does a new round begin?

2. What is the rate at which the appropriate nucleotides are incorporated into the new strand?

Cooper provided the essential clue for the answer to the first question by showing that the time required for a complete cycle of the fork was independent of the doubling time of the culture. Therefore, the increase in the DNA synthesis rate with increased growth rate can only occur because the frequency of a new cycle of replication has increased. Furthermore, this means that the frequency and growth rate are linked, since replication must occur only once for each cell division.

If m is the cell mass and n is the number of origins, then

$$m/n = k \qquad \text{a constant}$$

is the necessary condition; that is, when $m/n = k_1$, initiation will occur, thereby doubling the value of n. No further initiation can now occur until cell division has occurred and the mass has doubled. Then m/n will have returned to the former value of k_1 and a new cycle can occur.

If k is a constant, then the average amount of DNA per cell, expressed as equivalent chromosomes, is

$$\bar{G}/\bar{M} = \frac{\tau}{kC \ln 2} (1 - 2^{-C/\tau})$$

If C is constant, then $\overline{G/M}$ should decline with increasing growth rate, that is, smaller τ's; this is what Schaecter observed. For $\tau \leqslant 70$ min, $C \approx 40$ to 45 min.

In contrast to the pure DNA loop found in bacteria, the DNA in eucaryotes, those cells in which the genetic apparatus is localized in a membrane-defined nucleus, occurs as a set of complicated structures in which the DNA forms a complex with large amounts of protein. These structures are the chromosomes and they can be observed with optical microscopy during cell division. Although the amount of protein in the chromosomes is much larger than the amount of DNA, all evidence indicates that a single helix of DNA is the core of each chromosome. One can appreciate the structural complexity that must characterize the DNA in such an arrangement from the fact that a single DNA molecule in one of the chromosomes of the fruit fly (*Drosophila melanogaster*) proves to be linear and unbranched, but when extended is about 1.2 cm long. A similar result is found for human cells. The total DNA content of a human cell is about 10^{10} base pairs, which is equivalent to a total length of about 4 m. This DNA is packed in 46 chromosomes, but the total length of the chromosome is only about 200 μm.

Other distinctions between eucaryotic and procaryotic DNA are known. For example, if DNA is denatured by heating it to a temperature $> T_{melt}$, the renaturation can be followed by measuring the optical absorption at 260 nm, because single-strand DNA is about 40% more absorptive than double-strand DNA at this wavelength. From the previous equations, it is easy to see that the fraction of single-strand DNA will decrease with time during the renaturation according to

$$\frac{1}{1 + k_1 t [\text{DNA}]}$$

where [DNA] is the total concentration in moles per liter of nucleotides. Thus we expect the time for a given renaturated fraction to occur to depend on the reciprocal concentration. This is confirmed by experiments with sheared fragments of *E. coli* DNA; as the concentration of complementary fragments falls, the rate of renaturation decreases. This suggests that the speed of renaturation will increase as the abundance of complementary regions increases. The renaturation time for *E. coli* DNA is about 1 msec. Since mammalian DNA is about 10^3 to 10^4 times as large, one would expect such DNA to have a renaturation time of $\sim10^3$ to 10^4 msec. Hence, if the concentration of mammalian DNA is, for example, $\sim10^{-4}$ M, it will take about 10^8 seconds to attain 50% recombination. However, the actual time required for some of this DNA is in fact only seconds. The only reasonable explanation is that much mammalian DNA, in contrast to *E. coli* DNA, is characterized by a great many identical sequences. For example, from these arguments, it appears that about 10% of mouse DNA is simply a million copies of the same set of about 300 base pairs, while about 25% is roughly 104 copies of similar sequences. About 65% of the mouse DNA is "normal," that is, it is not composed of multiple copies. The function of such highly repetitive DNA is not known; however, there are two clearly important clues. One is that such DNA appears to be localized, when it is folded up, at the joint or centromere of the chromosome; the other is that the repetitive sequences are distributed along the DNA strand and not all clumped in one region.

Another intriguing distinction between eucaryotic and procaryotic DNA is the occurrence of long pallindromes, base sequences in double-strand DNA that would be seen as identical by the transcription process. These are common in the eucaryotes but unknown in the procaryotes. An example of a series of bases that form a palindrome is shown in Figure 5.5. Each strand of a pallindrome will produce an identical RNA molecule regardless of the point of origin.

Experiment shows that if double-strand DNA is separated by denaturing and then normal conditions are carefully restored, the pallindromic region forms a hairpin shaped loop on each strand, which can then be separated from the nonpallindrome

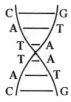

Figure 5.5 An example of a DNA palindrome.

DNA. Such structures can be as long as $\sim 10^3$ base pairs. Because these structures are common in DNA from a variety of sources and are generally distributed along the strands, and since separation of strands is a normal part of DNA function, it has been assumed that the pallindromes play some as yet unknown role in chromosome function. Heinje and Blomberg have investigated the dynamics of this process. Their results seem to show that whatever function the hairpin loops may be eventually found to serve, the structures themselves are relatively short-lived. Using a sophisticated random-walk calculation, which emphasizes that a recombination of single strands strongly favors the helix form, Heinje and Blomberg find lifetimes for the structure of < 1 second. This result is supported by the experiments of Wilson and Thomas, who were unable to detect stable hairpin loops from each strand in isolated double-strand DNA. Of course, these results do not say that such structures do not occur in native DNA under cellular conditions, but only that if they do, some additional stabilizing mechanism, perhaps the binding of a protein, is needed if the structures are to exist for times greater than about a second.

As the above remarks demonstrate, there are aspects of the genetic apparatus that remain to be understood. However, the general features of replication and protein synthesis are understood. One of the most recent advances is an explanation for the relatively low error rate of these processes. Of course, neither replication nor protein synthesis is an error-free process and, for example, the small changes in a particular protein from different species is presumably due to the incorporation of such errors. However, the processes associated with protein synthesis and DNA do show an error rate that is much smaller than would be predicted on the basis of probability arguments. In the case of protein synthesis, the error, that is, the rate at which the wrong amino acid is incorporated, is about 1 per every 3×10^3 amino acid choices; for DNA the rate is much lower, about 1 in 10^8. Of course, the large number of molecules involved in biological processes must be correctly produced; thus these low rates are necessary for the proper functioning of biological systems.

Suppose we have two products. If one is correct and the other wrong, the error rate f, obviously less than 1, will be controlled by the difference in the free energy, ΔF, between the two products:

$$f = \exp(-\Delta F/RT)$$

Hopfield has proposed that a process that he calls kinetic proofreading can substantially reduce the value of f. In his scheme, consider an enzyme that reacts with a substrate S, producing a product P. Consider also the possibility that the enzyme can react with a wrong substrate W, to produce an incorrect product. Hopfield suggests that the path of the correct enzymatic reaction is as shown in Figure 5.6. If

$$[S] = [W]$$

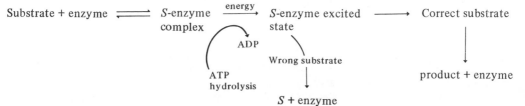

Figure 5.6 This drawing indicates schematically the process of kinetic proofreading.

then the decay scheme will be sharply in favor of P, and any W will be excluded by an amount at least $\approx f^2$. Hopfield's scheme correctly predicts the action of the enzyme isoleucyl-tRNA synthetase in discriminating between the correct incorporation of isoleucine and the incorrect choice of valine. The turnover at the excited state is the key step in the scheme. When the substrate is correct, the flow is directly to the product; when not, the flow is mostly a decay from the excited state to a form of the original substrate. Therefore, the energy cycle is very active for the wrong choice, but not for the right one. Since to get the energy requires the hydrolysis of ATP, one would expect low hydrolysis rates for the correct choice, but high rates for the wrong one. In the tRNA case, 1.5 ATP molecules are hydrolized per molecule of isoleucine correctly chosen, but 270 molecules of ATP per incorrect valine. Therefore, the error rate is reduced by 270/1.5 or a factor of 180, that is, from about 1 per 100 to about 1 per 18,000. This process is presumably the explanation of Berg's observation that in tRNAs, incorrect results are eliminated by hydrolytic attack of the enzyme.

It is now common knowledge that the information coded in the DNA base-pair order determines the characteristics that an organism inherits from its predecessor; however, it is possible to ascertain the rules by which such characteristics are transmitted and how they are distributed among the descendents without any knowledge at all of the role of DNA. In fact this was done in classical genetics through the patient observation of the results of a great many different cross-breeding experiments, particularly those involving the fruit fly, *Drosophila melanogaster*.

Of course it is impossible here even to summarize the essentials of classical genetics. Recognizing that, we can say that all of these experiments were understandable if one assumed that "heredity is particulate in nature and that the particles follow specific rules of transmission from parent to progeny." These particles have been given the name genes.

Through an immensely detailed study of heredity in fruit flies, T. H. Morgan was able to deduce the order of the genes and their relative spacing on the chromosomes, and these results were often given as chromosome maps. However, the association of the genes with the chromosomes remained somewhat tentative. Eventually, this association was confirmed by showing that in the case of broken chromosomes, there was a correlation between the damaged region and an altered or absent hereditary characteristic.

The next great advance came with the application of these methods to microorganisms. The traits studied were no longer just the appearance of the organism, but instead chemical characteristics such as a requirement or nonrequirement for a particular compound. These studies were begun by Beadle using the mould *Neurospora.*

Based on studies of genetically determined nutritionally defective cells, Beadle and Tatum showed that a particular deficiency was always related to a need for a single organic compound. They explained this result as the need for a substance, which would be produced in a normal cell but not in a defective one (i.e., a mutant), owing to the absence of an enzyme in the defective cell that could catalyze the necessary reaction. Further work accumulated a large number of cases in which specific enzymes could be shown to be present in normal cells but absent in mutants. From these studies, Beadle and Tatum concluded that the chemical basin of gene action was the specification of a particular enzyme by a particular gene, a result often shortened to "one gene — one enzyme."

Let us now look at the problem from the molecular end. In DNA, there are four letters in the code, the four possible bases, and in a protein there are twenty, the amino acids. Clearly, each nucleotide cannot specify just one amino acid, since that would give only four possible amino acids, not the twenty needed. Similarly, two nucleotides will not work because there are still only 16 combinations. Thus an amino acid must be specified by at least three bases.

The next step is to specify the code. The results are that each amino acid is specified by several different triplet bases. This is not really odd. If we use a triplet code, there will be 64 combinations, but only 20 amino acids to be specified. Furthermore, the order is important in that ACU is not UCA. Finally some of the triplets act to terminate amino acid chains and, when expressed on the mRNA, are not read by tRNA but by a protein known as a release factor. (Figure 5.7 summarizes the above information.)

Although we can by no means give a complete explanation, the outline of the connection between the molecular processes and the characteristics of the whole organism begin to appear. For example, an albino animal fails to make body pigment(s). Each reaction that would normally occur must be catalyzed by an enzyme. Therefore, albinoism may result from the malfunction or total absence of a needed protein. That molecular defect thus manifests itself as an observable characteristic. Many other examples, such as sickle-cell anemia, are known. Of course, to be able to understand completely how the information coded at the molecular level (the genotype) determines the characteristics of the organism (the phenotype), it will be necessary to understand the problems of development and differentiation of cells. All of these areas, from a study of the dynamics of the chromosomes to the problems of differentiation, are increasingly the subject of study from both biochemical and biophysical viewpoints.

To conclude these brief remarks on genetics, let us bring the classical concept of the gene into a molecular picture. From what has been said, it is clear that the gene is a portion of a DNA strand. It is important to remember that DNA is never used to

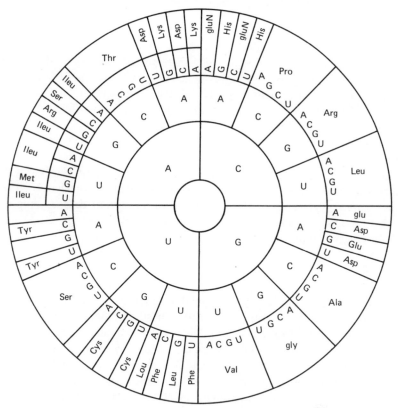

Figure 5.7 The above chart shows the various base triplets that code specific amino acids.

produce proteins directly, but always acts through transcription, the production of a complementary copy of a portion of one chain of DNA, the mRNA. This process is of course mediated by an enzyme, RNA polymerase. How big is a gene? One way to estimate this is to appeal to the one-gene/one-enzyme rule. If the average molecular weight for a single chain is 50,000, we need about 300 amino acids. Therefore, about 900 nucleotide pairs on the original double helix are needed to specify these amino acids. Since it is .34 nm between adjacent nucleotide pairs, the gene must be about 300 nm long. Of course in a eucaryotic cell, the DNA is complexed with protein and folded up as the structure of a particular chromosome. It is difficult to be sure just how large such a folded structure should be, but estimates of about 10 nm as a characteristic diameter of the folded form are common. In fact, by using radiation to induce a mutation, it can be shown (see Chapter 10) that the sensitive volume of a chromosome corresponding to a single gene would be a sphere about 5 to 10 nm in diameter.

How do these structural genes function? Although many questions remain, a very impressive model has been developed from the work of Jacob and Monod. We consider only the bacterial case. The genes can be divided into two categories.

First there are nonregulated genes, which code for those proteins that the cell must have, independent of its environment. These proteins are produced at a constant rate through the binding of an RNA polymerase to a short stretch of DNA adjacent to the gene. This short stretch of DNA is called a promoter, and the rates for the production of various proteins are determined by the relative affinities between the various promotors and RNA polymerase.

In contrast to the above are the regulated genes, which code for the production of proteins needed only under certain conditions. These genes are sensitive to the presence or absence of key molecules in the processes for which they code the enzymes. Regulated genes often occur together if they are functionally related and are often transcribed together onto a single mRNA. In procaryotic cells, a group of such genes that transcribe together is called an operon and can be as large as 20 genes. The model for operon control depends on three elements. First there is an operator gene that can bind the protein that will act as a regulator; second is a regulator gene to produce that protein; and third is a promotor to which RNA polymerase can bind in order to start the operon transcription. Consider one mode of action of the model, that in which the protein produced by the regulator gene acts as a repressor. Suppose we want enzyme production because a particular substrate is now abundant. Some of these molecules will bind to the protein. Since it is now bound, it cannot also bind to the operator. Then, since RNA polymerase can, mRNA synthesis is initiated with the consequence that enzyme production begins. The reverse is also possible; the protein (combined with some other molecule) is bound to the operator, blocking the binding of the polymerase and thus preventing enzyme production. These very condensed comments by no means exhaust the possibilities of this model, which is discussed in great detail in genetics texts.

chap. 6

The Cell Mem-brane

As was observed earlier, one of the characteristics of living matter as exemplified by a single cell was that the cell was enveloped by a membrane — properly, the plasma membrane — which was a dynamic component of the cell. Membrane structures are, in fact, involved at every level of cellular activity. In eucaryotic cells, membranes pervade the cell, occurring in mitochondria, the Golgi apparatus, the nucleus, and so on. In this chapter, we will consider only certain transmembrane properties of the plasma membrane, ignoring the many other roles that it plays in such matters as cellular contact, the interaction of hormones, or the immune response.

Clearly membranes, enzymes, and nucleic acids are all central to the processes of living matter. However, in the case of the enzymes and the nucleic acids, we have also been able to regard them as just complicated macromolecules. In fact, an example of each has been synthesized, although it is true that in the case of the nucleic acid, we need a "primer" strand that must be made by the cell. Nevertheless, if we look on the cell as a particularly handy laboratory for doing a bit of synthesis for us, we can more or less regard enzymes and nucleic acids as objects of study capable of existence and activity independent of any explicit association with living matter.

This is not the case for the next component of living matter that we wish to discuss, the membrane. There are, as yet, no membranes that can be synthesized in the laboratory and that will then perform as do those membranes produced by cells, nor can natural membranes long retain their functional properties if separated from the cellular context. There is no reason to believe that we cannot learn to make proper membranes, and indeed there is considerable progress in that direction, but at present the study of biological membranes has not progressed far enough to permit such a success. The membrane problem is thus one of the central problems of biophysics.

BASIC MEMBRANE PROPERTIES

The first task in the study of membranes is the determination of their composition. This is most conveniently done by obtaining membranes from cells that have been ruptured by ultrasonic, mechanical, or hypotonic means. If the fragmented cell material is now centrifuged, the various cell components will separate because of their different densities and will form a set of layers in the centrifuge tube. If these different layers are examined by means of electron microscopy, the particular layer containing the membrane fragments may be identified; the characteristic "railroad track" of membrane structure in electron micrographs is shown in Figure 6.1. This membrane fraction may then be analyzed by standard chemical techniques, by electrophoresis and by other methods, in order to determine the composition.

The membrane is always found to consist of three major components: proteins, lipids, and water. The lipids are usually phospholipids, but glycolipids and neutral lipids, mainly cholesterol, are also found to occur. These three components are present in varying amounts depending on the biological material from which the membrane was obtained. As the data in Table 6.1 shows, the percentage of protein and lipid is

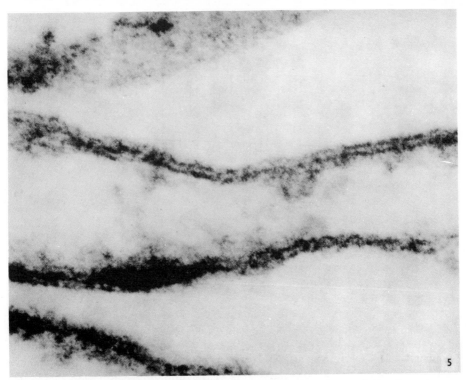

Figure 6.1 This electromicrograph shows the characteristic bilayer structure of the cell membrane.

variable; in general, the ratio of protein to lipid is between $\frac{1}{4}$ and 4. Furthermore, the particular form of the lipid component varies with the source; for example, cholesterol never occurs in bacteria such as *E. coli* and *Azobacter agilis*.

How are the macromolecules arranged in the membrane? One of the first clues to the molecular arrangement in membranes was provided in 1925 by Gorter and

Table 6.1 Membrane Composition

Source	Percent Protein	Percent Lipid
Rat liver	85	10
Rat muscle	65	15
Nerve membrane from ox brain	20	75
Nerve membrane from squid axon	18	80
Avian red blood cell	89	4
Human red blood cell	53	47
Bacteria:		
Micrococcus lysodeikticus	68	23
Bacillus megaterium	70	25

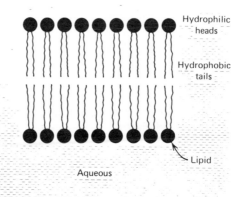

Figure 6.2 The schematic arrangement of lipid molecules in a bilayer, first proposed by Gorter and Grendel.

Grendel's studies of lipids extracted from red blood cells. In this work, they used techniques that had been developed by Langmuir in 1917 to produce and study monomolecular layers. In Langmuir's experiments, a very small drop of a liquid under study is placed on a very clean water surface. The drop spreads into a monomolecular layer, and a movable barrier is used to control the flow of the film.

Gorter and Grendel prepared such monolayers from the red blood cell lipids and showed that the area of the layer was nearly twice the surface area of the red blood cells from which the lipids had been extracted. This suggested that the natural arrangement of the red blood cell membrane was a double layer, or bilayer, of lipid, into or around which the protein was somehow arranged (Figure 6.2). The known size of the lipid molecules suggested that the membrane should be some 5 to 6 nm thick.

This model was attractive for two reasons. The first was that Overton had pointed out in 1899 that, in many cases, the relative permeability of the cell to many substances was directly proportional to the lipid solubility of those substances. The second argument in favor of the lipid bilayer model came from Fricke's experiments on the electrical properties of red blood cells. He found that at low frequencies, red blood cells are essentially nonconducting, but that the impedance of the suspension declined with increasing frequency. Fricke showed that these results would be obtained if the red blood cell was taken to be a conducting region surrounded by a thin insulating layer. For volume concentrations, ρ, smaller than about 20%, the capacitance, C, of a suspension of n red blood cells of radius r is

$$C = 9/4 \; r\rho C_s$$

where C_s is the capacitance of a single cell. Fricke consistently found

$$C_s \approx 1\mu F/cm^2$$

Since

$$C_s = \frac{\epsilon}{4\pi\Delta x}$$

where ϵ is the dielectric constant and Δx the layer thickness, it follows that for reasonable values of ϵ

$$4\,nm \leqslant x \leqslant 10\,nm$$

Further support for the bilayer model came from optical birefringence studies, which indicated that the lipids were perpendicular to the cell surface and to the protein component.

These results, and several similar studies, led Davson and Danielli to suggest a fairly specific membrane model (Figure 6.3) which, in a sense, became the model against which new ideas were judged.

Arguments about the detailed structure of membranes are often tied to the properties of the membrane. Before continuing the discussion of membrane structure, let us discuss the properties of the membrane that any model must somehow explain.

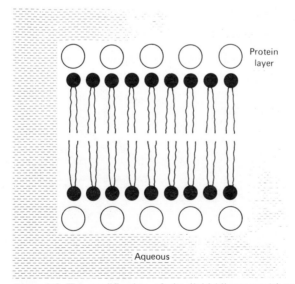

Protein layer

Aqueous

Figure 6.3 The modification of the lipid bilayer model to include protein, as proposed by Dawson and Danielli.

ANALYZING TRANSPORT

One of the main membrane properties is to serve as a barrier, not only between the various contents of the cell and its surroundings but also between the various components or compartments inside the cell. The first problem is how to determine the effectiveness of this barrier against the passage of various substances. We begin by defining the effectiveness of the barrier in terms of the amount of a particular substance that crosses the barrier per unit time; this amount is the flux, J.

One of the most useful methods for measuring the flux is to employ isotopic tracers. There are several techniques for carrying out such a determination. A common one is the wash-out experiment, in which the cells are first exposed to a bath containing the tracer isotope. The labelled material is taken up and, after a time long enough to be confident that equilibrium has been reached, the cells are transferred to an unlabelled solution. The amount of tracer in the cells will now change with time according to

$$\frac{da_2}{dt} = -J_{21}\pi_2^* + J_{12}\pi_1^*$$

where the specific activity π^* is the ratio of the amount of labelled substance to the total amount of substance.

Since we began in equilibrium

$$J_{12} = J_{21}$$

The volume of the unlabelled solution to which the cells were transferred is assumed to be large compared to the cell volume. Therefore,

$$\pi_1^* \simeq 0$$

and we have

$$\frac{da_2}{dt} = -J_{21}\pi_2^*$$

Using the specific expression for π_2^*, we have

$$\frac{da_2}{dt} = -\frac{J_{21}}{S_2}\,dt$$

Integrating,

$$\ln \frac{a_2}{a_{2(t=0)}} = \frac{-J_{21}}{S_2} t$$

where a_{20} is a_2 at $t = 0$. Therefore,

$$\ln a_2 = -\frac{J_{21}}{S_2} t + \ln a_{20}$$

The rate at which the fraction S flows from the compartment is the rate constant, k:

$$k \equiv J/S$$

The observational data is the number of counts per minute as a function of the time. If this is plotted, the slope is k. If we then measure the amount in the cell S_2, which can be done by spectroscopy or other means, we can compute the flux J_{21}.

Furthermore, since the volume of the compartment is

$$v = a_{20}/\pi_{20}$$

we can also calculate v by taking the y-intercept on the plot of $\ln P_2$ versus time.

The measurement of the fluxes of various substances across cell membranes produces a variety of results, and we are not yet able to explain all of them. Let us now consider how the motion or transport of materials across the membrane might be described from a theoretical viewpoint.

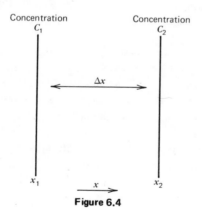

Figure 6.4

The simplest process that could lead to the movement of a substance across the membrane is diffusion. The one-dimensional case is shown in Figure 6.4. Suppose the concentration of the diffusing substance is uniform across the plane at x_1 and increases in the direction of $x = 0$. The probability that a molecule at some point x will move toward or away from the direction of increasing x is equal. However, there are more molecules in the direction of decreasing x. Therefore, many more molecules will move in the direction of increasing x than in the opposite direction, and the substance will diffuse from the region of higher concentration toward the region of lower concentration. Thus

$$\frac{\Delta m}{\Delta t} = \beta(c_1 - c_2)$$

so

$$\frac{\Delta m}{\Delta t} = \text{constant} \cdot \frac{\Delta c}{\Delta x}$$

The constant is called the diffusion constant and the equation is known as Fick's diffusion equation or Fick's law.

The mass contained in the volume between x_1 and x_2 will change with time; this rate will be the difference between the rate of accumulation through the area at x_1 and the rate of loss through the area at x_2. Therefore,

$$\Delta(\partial m/\partial t) = D(\partial c_2/\partial x - \partial c_1/\partial x) => \frac{\partial c}{\partial t} = D\frac{\partial^2 c}{\partial x^2}$$

Thus, to describe diffusion through a material, we can either determine D and predict results or use the measurements of the flux to deduce the value of D appropriate to the material.

Numerous experiments show that many substances to which the cell membrane is permeable clearly do not obey the simple diffusion equation. Indeed, examples are known in which the fluxes are precisely the opposite of what the diffusion equation would lead one to expect. For example, potassium ions are accumulated in the cell when the concentration of potassium in the cell may already be a hundred times the concentration outside the cell.

The obvious inability of the diffusion equation to describe membrane transport has led to efforts to derive other analytical methods. Three general approaches have been suggested. The first of these is based on the Nerst-Planck equation, which forms the basis for investigating motion under the influence of electric potential difference, or electrodiffusion. The second approach is based on the formalism of irreversible thermodynamics, while the third takes its origins from Eyring's theory of absolute reaction rates. Before considering these three, it is well to point out that none of these methods provide us with all the answers.

Approaches based on the Nernst-Planck equation or on irreversible thermodynamics have in common derivations that are concerned with the forces thought to be present in the system. In the Nernst-Planck approach, we begin by observing that the flux should be given by an expression of the form:

$$\text{Flux} \equiv \text{mobility} \times \text{concentration} \times \text{force}$$

or

$$J = \omega c F$$

where ω is the molar mobility per unit force. What forces may be present? Since any force can be represented as the negative gradient of a potential:

$$F = -\text{grad } \mu$$

we can either specify the force directly or describe it in terms of the appropriate potential gradient:

1. A force due to the change in chemical potential across the membrane. Since most substances are assumed free to move in the various cell compartments and are present in low concentration with respect to water, this force becomes the concentration gradient, dc/dx.

2. Electrical potential gradients, $d\psi/dx$

3. Pressure gradients, dP/dx.

Therefore, after transforming so that all quantities are in the same units, the total possible force is

$$\frac{RT}{c}\frac{dc}{dx} + \bar{v}\frac{dP}{dx} + z\mathscr{F}\frac{d\psi}{dx}$$

where $\mathscr{F}$ is the Faraday and $\bar{v}$ is the specific volume. This is taken as negative and set equal to the negative gradient of a potential that is introduced to represent the net force and is called the electrochemical potential. Thus

$$J = -\omega c RT \left(\frac{1}{c} \frac{dc}{dx} + \frac{\bar{v}}{RT} \frac{dP}{dx} + \frac{z\mathscr{F}}{RT} \frac{d\psi}{dx} \right)$$

Obviously, to predict the flux from this equation requires no little knowledge of the conditions under which membrane transport occurs.

In order to demonstrate that the above equation is at least reasonable, let us show that Fick's law of diffusion can be obtained from it. Assume a neutral substance and the absence of any pressure gradient across the membrane. Then the above equation becomes

$$J = -\omega RT \, dc/dx$$

but

$$D = \omega RT$$

so

$$J = -D \, dc/dx$$

or

$$J \, dx = -D \, dc$$

The flux is independent of the thickness of the membrane; the thicker the membrane, the longer you wait, but the flux is the same. If we assume the concentration across the membrane is linear, then

$$\int J \, dx = -\int D \, dc$$

so

$$J \int dx = -D \int dc$$

or

$$J(x_2 - x_1) = -D(c_2 - c_1)$$

Now the flux is the amount of material per unit time:

$$J = \Delta m / \Delta t$$

so

$$\Delta m / \Delta t = D \Delta c / \Delta x$$

which is just Fick's law.

The second approach, based on Onsager's formulation of irreversible thermo-dynamics, is more subtle because it permits us to deal with the possibility that all of the fluxes are related to all of the forces acting.

The central assumption in this approach is the division of the entropy change into two parts, one due to the internal processes of the cell and the other, to the external processes. For the part due to internal processes, we can write

$$\frac{d_i S}{dt} = -\frac{1}{T}\frac{dG}{dt}$$

From this it can be shown that

$$T \, di \, S/dt = \sum_i J_i X_i$$

where J is the flux and X is the force. We now write the relation between the fluxes and the forces acting to produce them as

$$J_i = \sum_{j=1}^{N} L_{ij} X_j$$

That is,

$$J_1 = L_{11}X_1 + L_{12}X_2 + \ldots$$
$$J_2 = L_{21}X_1 + L_{22}X_2 + \ldots$$
$$\vdots$$
$$J_n = L_{n1}X_1 + L_{n2}X_2 + \ldots$$

where the coefficients L_{ij} link the fluxes and forces. Thus a particular flux can be considered as due to an appropriate force, sometimes called the conjugate force, plus contributions from all the other forces acting in the system. Hence

$$J_i = L_i \Delta \mu_i + \sum_{j \neq i} L_{ij} \Delta \mu_j$$

where $\Delta \mu_j$ is the electrochemical potential. We can normally assume that the major source of a given flux is its conjugate force.

Again, we can demonstrate the reasonableness of this approach with two examples. First, we show that the above approach accurately produces Fick's law, and then we treat a purely physical case, the thermoelectric effect, which demonstrates the correctness of the above scheme for a situation where two fluxes and two forces are acting. Let us now consider the diffusion problem. We begin by writing

$$\Delta \mu = RT \ln c_1 / c_2$$

so

$$J = L_{ii} RT \ln c_1 / c_2$$

but

$$\ln x = 2(x - 1/x + 1) + \ldots$$

therefore

$$J = 2 L_{ii} RT \frac{c_1/c_2 - 1}{c_1/c_2 + 1} = \text{constant } (c_1 - c_2)$$

which is Fick's law.

The idea that all the forces and fluxes acting in a system are somehow linked may seem at first a bit strange. It is, therefore, worth pointing out that experimental results from studies of very well-defined and absolutely nonliving systems completely support that idea. As an example, consider thermoelectricity.

In 1821, Seebeck was investigating the thermal properties of junctions of copper and iron strips, arranged as shown in Figure 6.5. He discovered that if one of the junctions was heated while the other was cooled, a small potential drop, about 1 mV for $\Delta T = 50 °C$, was produced. In 1833, Peltier discovered the "opposite" effect: If a current flows in the circuit, one junction is heated and the other is cooled. Furthermore, the effect is reversible, in that upon reversing the direction of the

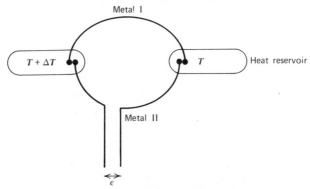

Figure 6.5 The thermoelectric effect.

current I, the junction that was cooled is now heated, and vice versa. The effect is clearly not Joule heating, which is due to resistance. If it were, it would be independent of current direction and would depend on I^2, whereas the heat generated or lost at a particular junction is proportional to I.

From the viewpoint of irreversible thermodynamics, there are two fluxes — heat and current — and two forces — electrical potential and thermal gradient. Thus we write

$$J_{\text{heat}} = L_{11}\left(-\frac{1}{T}\frac{dT}{dx}\right) + L_{12}\left(-\frac{d\phi}{dx}\right)$$

and

$$I = L_{12}\left(-\frac{1}{T}\frac{dT}{dx}\right) + L_{22}\left(-\frac{d\phi}{dx}\right)$$

where T is the temperature, ϕ is the emf, I is the current, and the L's are the coupling constants. From the conditions of Seebeck's experiment, $I = 0$ and so we can write

$$\left(\frac{d\phi}{dT}\right)_{I=0} = \frac{L_{21}}{L_{22}}\frac{1}{T}$$

Similarly for Peltier's experiment,

$$\left(\frac{J_{\text{heat}}}{I}\right)_{dT=0} = \frac{L_{12}}{L_{22}}$$

Since by Onsager's argument, we expect that

$$L_{21} = L_{12}$$

we have

$$(J_{\text{heat}}/I)_{dT=0}/T(d\phi/dx) = -1$$

Miller has assembled data for a variety of metals in order to test this prediction and finds it verified.

The third approach comes from solid-state physics and is an application of absolute reaction-rate theory. In this view, we picture a molecule of the substance under consideration as progressing through the molecular array of the membrane by a series of discrete jumps. The rate of jumps in the correct direction is

$$\Gamma = \nu e^{-E/kt}$$

where E is the energy needed to make the jump and ν is the average frequency of vibration of the points making up the structure through which the substance is transported.

Suppose the membrane contains n_1 such particles being transported. Then the concentration c_1 is given by

$$c_1 = n_1/\lambda$$

where λ is the thickness of the membrane. Therefore, the flux from x_1 to x_2 will be

$$J_{12} = \Gamma\theta_n = \Gamma\theta\lambda c_1$$

where θ is the fraction moving from $1 \to 2$. Thus the net flux will be

$$J = \Gamma\theta\lambda(c_1 - c_2)$$

Since the membrane is thin,

$$\frac{c_1 - c_2}{\lambda} \doteq -dc/dx$$

and therefore,

$$J = -\Gamma\theta\lambda^2 \, dc/dx$$

This result obviously contains Fick's law.

ELECTRODIFFUSION

Another fundamental property of the cell membrane is the potential drop that exists across the membrane. These electrical potentials can be measured with micro-electrodes, fine glass tubes filled with a conducting solution such as potassium chlorides. The bore of the microelectrode is so small that capillary attraction prevents any leakage and also insures that the electrode has a very high impedance, of the order of megaohms, so that no current is drawn during the potential measurement. The tip diameter of the microelectrode may be $\sim\frac{1}{2}\mu$ and is thus fine enough so that it presumably does not damage the cell into which it is inserted. Since the functions of such cells seem to be unaltered after such measurements have been made, the observed potentials are almost certainly those characteristic of the cell. Measured potentials are usually on the order of 50 to 150 mV, with the inside of the cell at negative potential. In addition to these relatively stable values, we also find some membranes in which the potential changes dynamically in response to a stimulus, and such changes continue for a time after the stimulus is removed. Such membranes are called excitable; an obvious example is the membrane around a nerve fiber, but excitable membranes are also found in other cells, such as certain algae.

The existence of a potential difference across a membrane through which ions must move immediately suggests an analysis from the viewpoint of electrodiffusion theory. The force on an ion will be

$$F = zeE$$

Frictional forces will quickly establish equilibrium. Assuming Stokes' law, the velocity of the ion will be given by

$$v = \frac{ZeE}{f} \doteq \frac{zEe}{6\pi\eta r} \times \frac{N}{N} = \frac{zFE}{N6\pi\eta r}$$

Thus

$$J = c \times \frac{zFE}{N6\pi\eta r} = -cUzF\frac{\partial\psi}{\partial x}$$

If we add the effect of diffusion, we obtain

$$J = -cU \operatorname{grad} (\mu + zF\psi) = -cU \operatorname{grad} \bar{\mu}$$

where

$$\mu = \mu_0 + RT \ln c$$

So

$$
\begin{aligned}
J &= -cU \left(RT \frac{\partial \ln c}{\partial x} + zF \frac{\partial \psi}{\partial x} \right) \\
&= -RTU \frac{\partial c}{\partial x} - zFcU \frac{\partial \psi}{\partial x}
\end{aligned}
$$

A solution to this equation was obtained by Planck, who assumed that

$$-\frac{\partial J}{\partial x} = \frac{\Delta c}{\Delta t}$$

where

$$c = \sum_{j=1} c_j^{+} = \sum_{j=1} c_j^{-}$$

Then define

$$
\begin{aligned}
\bar{u} &= \sum u_j^{+} c_j^{+} \\
\bar{v} &= \sum u_j^{-} c_j^{-}
\end{aligned}
$$

Then it can be shown that

$$\frac{\partial \psi}{\partial x} = -\frac{RT}{F} \frac{\dfrac{\partial(\bar{u} - \bar{v})}{\partial x}}{\bar{u} + \bar{v}}$$

If we assume a smooth distribution of the ion across the membrane:

$$c = \frac{c_i - c_0}{\delta} x$$

we obtain for the case of a single ion the Nernst-Planck equation:

$$\psi_i - \psi_0 = \frac{RT}{F} \ln \frac{c_i}{c_0}$$

which is to be understood to mean that an equilibrium situation in which the ion concentration is such that $c_i \neq c_0$ will always be associated with a potential difference across the membrane. The central problem then becomes a question of how the various substances are distributed and how they move. This is the point at which a new complication enters; some substances can be shown to move across the membrane only because metabolic energy is expended to do the work of that transport. Thus, another property of the membrane is the existence of active transport. The active transport process has been extensively investigated, however, the actual details at the molecular level remain obscure. In such situations, it is often useful to appeal to thermodynamics, which treats such systems as "black-boxes" and makes no specific demands for a knowledge of molecular details. The advantage of the thermodynamic approach is that, among other things, it may reveal missing factors or show interrelations that we would not have suspected.

ACTIVE TRANSPORT IN RED BLOOD CELLS

Katchalsky has shown how some of the elementary aspects of active transport in the red blood cell can be analyzed by means of irreversible thermodynamics. As you probably know, the red blood cell originates by the differentiation of stem cells in the bone marrow; because of that process, red blood cells are without a nucleus and therefore incapable of reproduction. These especially simple cells provide us with positive evidence of the central role of membrane-centered processes. As you can easily show (see Appendix), it is possible to remove the normal cellular contents and replace them with simple solutions of proper osmolarity; such cells are called ghosts. They will continue to function and, in their essential cellular properties, are indistinguishable from normal cells, in that they continue to exhibit active transport and the required metabolic activity.

The most impressive manifestation of active transport by membranes is the fact that potassium ions are concentrated in the cell, while sodium ions are removed. At physiological temperature, this active process is very effective; if we define the selectivity coefficient as the ratio of [K]/[Na] inside the cell to that outside the cell, we find for red blood cells,

$$\frac{[K]^i/[N]^i}{[K]^o/[N]^o} \simeq 2 \times 10^2$$

It has been shown that the energy required to maintain this situation is $\sim 3\frac{1}{4}$ kcal per mole of ion transported. The source of this energy, as discussed in Chapter 9, is the hydrolysis of ATP to ADP and P_i. Experiment shows that:

1. Na and K are both needed for ATP hydrolysis to proceed at the maximum rate.

2. External Na-cations and external ATP are of no importance; only internal Na-cations and internal ATP are required.

From a variety of biochemical experiments, the two processes, ion transport and ATP hydrolysis, have been shown to be linked.

In irreversible thermodynamics, we analyze the problem in terms of flows or fluxes, and forces. We can identify the fluxes as:

1. J_{Na}

2. J_{K}

3. The flow of some anion, J_A, which maintains charge balance.

4. The flow of the chemical reaction J_r, which provides the energy, that is, the energy flow for active transport.

The forces, in the sense of irreversible thermodynamics, which are related to these fluxes are:

1. The differences in the electrochemical potentials, measured across the membrane, for each of the ions.

2. The affinity of the chemical reaction that provides the energy.

Specifically, these may be written:

1. $\Delta\mu_{Na} = \mu_{Na}^i - \mu_{Na}^o$

2. $\Delta\mu_K = \mu_u{}^i - \mu_K{}^o$

3. $\Delta\mu_{An} = \mu_{An}{}^i - \mu_{An}{}^o$

4. $\mu_{ATP} + \mu_{H_2O} - \mu_{ADP} - \mu_{Pi} = A$

Now we know that

$$J_i = \Sigma L_{ij} X_j$$

or, equivalently,

$$X_j = \Sigma R_{ji} J_i$$

Therefore, we may write

$$\Delta\mu_{Na} = R_{Na} J_{Na} + R_{Na,K} J_K + R_{Na,An} J_{An} + R_{Na,r} J_r$$

$$\Delta\mu_K = R_{K,Na} J_{Na} + R_K J_K + R_{K,An} J_{An} + R_{K,r} J_r$$

$$\Delta\mu_{An} = R_{An,Na} J_{Na} + R_{An,K} J_K + R_{An} J_{An} + R_{An,r} J_r$$

$$A = R_{r,Na} J_{Na} + R_{r,K} J_K + R_{r,An} J_{An} + R_r J_r$$

A number of experiments indicate that the anion is not actively transported; its flow is purely passive and occurs only in response to the flow of the cation, as required for electrical balance. Therefore,

$$R_{An,r} = 0$$

Suppose we consider a red blood cell ghost that has been allowed to come to equilibrium. In that case,

$$J_{Na} = J_K = J_{An} = 0$$

Therefore, for static equilibrium, our set of equations can be written

$$\Delta\mu_{Na} = R_{Na,r} J_r$$

$$\Delta\mu_K = R_{K,r} J_r$$

$$\Delta\mu_{An} = 0$$

$$A = R_r J_r$$

Now if the solutions are reasonably ideal we may write

$$\Delta\mu_K + \Delta\mu_{An} \equiv \Delta\mu_{K,An} = R_{K,r}J_r$$

$$\Delta\mu_{Na} + \Delta\mu_{An} \equiv \Delta\mu_{Na,An} = R_{Na,r}J_r$$

$$\Delta\mu_{Na} - \Delta\mu_K \equiv \Delta\mu_{Na,K} = (R_{Na,r} - R_{K,r})J_r$$

But we know that

$$\Delta\mu_{Na} + \Delta\mu_{An} = RT \ln \frac{[Na]^i/[An]^i}{[Na]^\circ/[An]^\circ}$$

$$\Delta\mu_K + \Delta\mu_{An} = RT \ln \frac{[K]^i/[An]^i}{[K]^\circ/[An]^\circ}$$

$$\Delta\mu_{Na} - \Delta\mu_K = -RT \ln \frac{[K]^i/[Na]^\circ}{[K]^\circ/[Na]^\circ}$$

These quantities may be determined from experiment. We have

$$\Delta\mu_{Na} + \Delta\mu_K + 2\Delta\mu_{An} = 1.62 \text{ kcal}$$

for the case of Cl^- or HCO_3^-. Therefore,

$$R_{Na,r} = -R_{K,r}$$

This means that each K-ion is accumulated, and each Na-ion rejected in a fully coupled way.

Now according to Onsager's arguments,

$$L_{ij} = L_{ji}$$

so

$$R_{Na,r} = R_{r,Na}$$

We have

$$\Delta\mu_{Na,K} = R_{Na,r}J_r - R_{K,r}J_r$$
$$= R_{Na}J_{Na} + R_{Na,r}J_r$$

and

$$A = R_{r,Na}J_{Na} + R_rJ_r$$

But in equilibrium

$$J_{Na} = 0$$

$$\therefore \quad \frac{\Delta\mu_{Na,K}}{A} = \frac{R_{Na,r}}{R_r}$$

or

$$\Delta\mu_{Na,K} = RT \ln \frac{[K]^i/[Na]^i}{[K]^\circ/[Na]^\circ} = A \frac{R_{Na,r}}{R_r}$$

Thus, to keep the observed imbalance in cation concentration, even at equilibrium, the coupling between the ATP hydrolysis and Na-flux must remain finite. Furthermore, $\Delta\mu_{Na,K}$, which is the force responsible for separating the two cations, is seen to be temperature dependent. This is in fact the case; as the temperature falls below the physiological range, the ratio:

$$\frac{[K]^i/[Na]^i}{[K]^\circ/[Na]^\circ}$$

falls to one ($\Delta\mu_{Na,K} = 0$ at 0°). The ghosts are not harmed by this treatment provided that they are not actually frozen or held at a low temperature for a long period; even then some protection is afforded by treatment with glycerol. If the ghosts are now warmed, a flow of Na will occur until the ratio of Na/K is appropriate to the temperature. It is the dominance of the chemical reaction by the temperature that is the controlling factor. This situation is equivalent to making $A = 0$, as can be seen from the fact that

$$\left(\frac{J_{Na}}{J_r}\right)_{\Delta\mu_{Na,K}=0} = \frac{R_{Na,r}}{R_{Na}}$$

and

$$\left(\frac{J_r}{J_{Na}}\right)_{A=0} = \frac{R_{r,Na}}{R_r}$$

The previous calculation is obviously very general. However, if we choose to provide a specific model for the transport process, the same style of calculation can be applied and, of course, that model must produce predictions that are in agreement with the general results.

MEMBRANE MODELS

A very large number of experiments and theoretical studies, comprising far too much material to even mention, let alone discuss, have been focused on membranes, especiallly during the last 15 years. From the results of this work, it is now possible to outline a general model of the membrane. First, membranes are always formed by lipid bilayers, which are held together by interactions that are noncovalent and cooperative. Roughly speaking, these lipids represent half the mass of the membrane. Although a variety of membrane lipids is known, they share a common feature: All are amphipathic molecules, meaning that they have both hydrophilic (polar) and hydrophobid (nonpolar) parts. Such lipid bilayers are barriers to the flow of ions and most polar molecules, though water is an exception.

Protein Control of Permeability

Mechanism	Features	Experimental Tests
Carrier protein	Protein partially changes its conformation when substrate binds to it. This change permits the substrate to move across the membrane.	1. Transport reaches an upper limiting rate at high-substrate concentration. All carriers are involved and $\therefore$ no further rate increase is possible. 2. Transport of the normal substrate can be reduced by supplying a second molecule that is close enough in structure to be bound by the protein.
Molecular pump	Uses ATP hydrolysis to actively transport ions against the concentration gradients.	1. Na-K ATPase will hydrolize ATP only if K, Na, and Mg are present. 2. Ca^{++}-ATPase functions for Ca^{++}.

Since cell membranes can transport ions and polar molecules, it is a fairly reasonable conclusion that such transport properties are conferred on the lipid bilayer by some of the proteins associated with it. This is generally supported by the following observation: The principal function of nerve myelin membrane is to provide electrical insulation, so it should be relatively impermeable. Therefore, we would expect a high lipid abundance and a small protein component. On the other hand, since the plasma membrane is involved in a situation where transport is of critical importance, the protein component should be large. This is, in fact, what analysis shows: Myelin is only about 18% protein (presumably mainly structural), while the plasma membrane is about 50% protein.

The above conclusions can be considerably sharpened. Experiment shows that specific proteins give the membrane specific functional capabilities. This is particularly true for transport. For example, consider the red blood cell membrane, analyzed earlier by the methods of irreversible thermodynamics. The important characteristic of this membrane is its ability to concentrate K^+-ions while excluding Na^+-ions. In order to do this, energy must be provided through the hydrolysis of ATP. In 1957, Skou discovered an enzyme, called Na-K-ATPase, which hydrolizes ATP only if the ions Na^+, K^+, and Mg^{++} are present. It is commonly believed that this enzyme is the key feature of a molecular mechanism, called the Na-K pump, by which the transport of these ions through the membrane occurs. A similar enzyme is found in membranes associated with muscle tissue, but in this case the ion transported is Ca^{++}. Specific proteins that are not enzymes are also involved in other transport mechanisms for molecules. The most famous example of such a protein is the carrier of glucose in *E. coli*; in this case mutants of the bacterium are known that grow on maltose but not on glucose media. (Normal *E. coli* can do both.) It can be shown that there is nothing wrong with the metabolic utilization of glucose by the bacterium; hence the mutant is simply a cell in which the specific carrier protein for glucose is not present.

How are these proteins arranged in the membrane? A variety of experiments show that membranes are asymmetric. For example, consider the Na-K-ATPase: ATP is a substrate for the enzyme only when inside the red blood cell, and the enzyme functions only for situations in which Na^+ is inside and K^+ is outside. Thus, the protein complex of the pump must be an oriented structure in the membrane. In the case of red blood cells, experiments using labelled compounds and specific enzymes show that there are two major glycoproteins that extend from the external surface to the cytoplasma; however, most proteins are found on the cyctoplasmic side. As another example of asymmetry, sugar residues can be shown to occur only on the outside of mammalian cell membranes. The experimental fact that cell membranes are not completely symmetrical bilayers leads us to divide membrane proteins into two categories. The peripheral proteins are those that appear to be attached directly to the membrane surface; the integral proteins are those that have a substantial part of the molecule, in effect, buried in the bilayer. In the case of peripheral proteins, the binding forces are presumably covalent links and hydrogen bonds, since such proteins can be detached from the membrane by washing the cells with a salt solution. Integral proteins, on the other hand, are probably tangled with the lipid chains in the bilayer; only by treatment of the membrane with organic solvents or strong detergents can they be isolated. Of course, this classification should not be taken too literally as a specification of molecular arrangement, and you should be aware that the classification emphasizes the extremes of what is really a continuum. The importance of this work is that it clearly shows how the rather uniform protein layer of the original Davson-Danielli model must be altered to a nonuniform arrangement.

The other major evolution of the Davson-Danielli model came with the realization that the structural arrangement of lipid bilayer and protein could not be thought of as a particularly rigid system. Experimental results consistently point to the conclusion that at least a substantial portion of both the protein molecules and the individual lipid molecules are free to move in the plane of the membrane. NMR determinations of the diffusion constant for phospholipid molecules give values of $D \approx 10^{-8}$ cm^2/sec. This means that a particular lipid molecule will be displaced by lateral diffusion in the bilayer an average distance $\langle x \rangle$ in a time, t, given by

$$\langle x \rangle = \sqrt{4 Dt}$$

On the molecular scale, this means that a given lipid molecule can exchange positions with an adjacent lipid at a frequency of about 1 MHz. On a cellular scale, this produces an average lateral velocity, that is, a motion in the bilayer plane, of about 2 μm/sec. Since the average length of a bacterial cell is only a few micrometers, the fluid nature

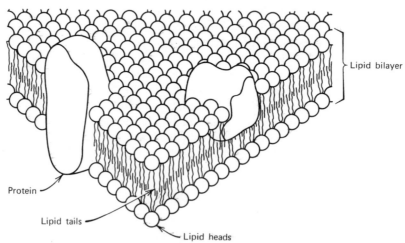

Figure 6.6 An example of a recent membrane structure model is the fluid mosaic model developed by Singer and extended by others. The model retains the lipid bilayer, but proteins may be attached either at the surface or through the lipid bilayer. Furthermore, at least some of the proteins are free to move laterally in the plane of the bilayer.

of the membrane is obvious. In light of this conclusion, it is worth recalling that Oventon's original discovery was that the permeability of the membrane to a number of different nonpolar molecules was proportional to their diffusion rates in olive oil, whose viscosity is only some 200 times that of water.

All of the above results have led to a general evolution of the Davson-Danielli model into the fluid-mosaic model, first proposed by Singer and Nicholson in 1970. The fluid-mosaic model retains the basic lipid bilayers; the role of the bilayer is now seen to be both a permeability barrier and a structural matrix for proteins. The proteins control the specific permeability of the membrane, can diffuse in the plane of the membrane, and may be embedded in one of the layers or may extend through the bilayer. Thy lipids have a somewhat freer motion, but in both cases the motion is in the plane of the membrane, as shown by the fact that lipid-lipid exchange between the two layers of the membrane occurs at a rate less than 1 per 1.3×10^6 sec. A general fluid-mosaic model is illustrated in Figure 6.6.

chap. 7

The Nerve Impulse

Our knowledge of our surroundings and our responses to that knowledge depend on the interactions of very highly specialized cells, the neurons. The nervous system, not only in human beings but also extending even to very simple animals, is the result of the activity of sets of neurons connected with varying degrees of complexity.

Neurons (Figure 7.1) are irregularly shaped cells. The cell body of the neuron is characterized by clumps of endoplasmic reticulum (Nissle bodies) and by a very large nucleus, which however contains the same amount of DNA as do other cells. The processes or projections of the cell body are the dendrites and axon. Typical neurons have many short, branched dendrites and a single long axon that may have a very few branches; neurons having only one dendrite and one axon also occur. Most axons have a diameter of 30 to 50 μm which is small compared to the size of the cell body. However, they can reach lengths of several meters in large animals; in human beings, the longest axons are about 1 m. Also the end of the axon is specialized; for example, a motor neuron axon terminates at a voluntary muscle with a specialized motor end plate. Nerves are collections of axons. The unit structure of most nerves is the funiculus, a bundle of axons covered with connective tissue and supplied with blood

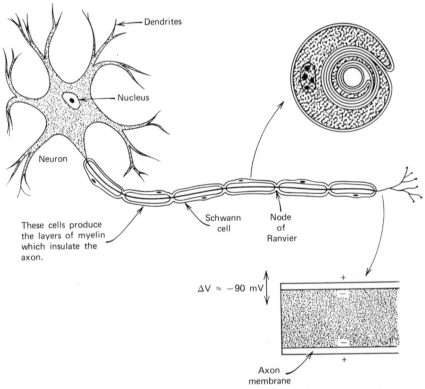

Figure 7.1 The above drawing shows the elementary structure of the neuron and axon.

vessels. A bundle of funiculi, covered by a second sheath of connective tissue makes up the nerve. One aspect of neuron specialization is that certain axons only carry signals toward the brain, while others only carry signals away from it. Although this distinction between afferent and efferent neurons is not physiologically insignificant, it is not critical for understanding the process by which an individual axon transmits the nervous impulse, since it has been shown that the axon, if correctly stimulated, is physically able to conduct a signal in either direction. This is not surprising since, as we will see, the mechanism by which the signal is transmitted is located in the axon membrane and depends on the flow of ions across the membrane but not along it.

Notice that although the conduction along a particular axon need not be in only one direction, there is another factor to be considered and it is this factor that makes the signal direction uniform for a particular axon. As you know, neurons are interconnected; the junctions are called synapses. As an example, consider an axon near a dendrite. Electron microscopy shows that the two are not in direct contact. In fact, at the synapse, the conduction of the signal depends on the release of a neurochemical transmitter, acetylcholine (ACh), from the end of the axon in response to the electrical signal. This substance diffuses across the synapse gap in about 1 msec where it acts to change the sodium permeability of the membrane and generate a new action potential. Roughly 10^3 ACh molecules per pulse are required. Immediately after the generation of the new signal, these ACh molecules are inactivated by the enzyme, ACholinesterase. It is this chemical transmission that fixes the actual direction a signal may propogate along the axon.

Our initial goal is to analyze the conduction of the nerve impulse along the axon. Microelectrode measurements, for example, on the optic nerve fibers in a horse-shoe crab, quickly reveal the essential facts about the nerve signal. The signal takes the form of a short-lived voltage pulse, often called the action potential (Figure 7.2). The frequency with which the signals occur is a function of the strength of the stimulus; in this case, as the illumination of the crab's eye increases in intensity, the number of

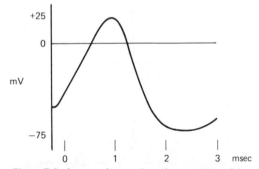

Figure 7.2 A general nerve impulse as observed in the axon is shown. The characteristic features are the rise from a negative potential, the decline to a value more negative than the resting potential, and the slow return to the resting potential.

pulses per second transmitted by the nerve increases. Furthermore, the amplitude of the pulse is independent of the strength of the stimulus. There is also a refractory period, a short time interval after the passage of a pulse, when the nerve cannot be stimulated to conduct another pulse. The velocity of the pulses along the fiber is found to be essentially a constant, independent of the direction the pulse is traveling. It is also possible to produce signals by directly stimulating the nerve with a short-voltage pulse. This experiment, as well as other results from different axons, shows that a minimum voltage is necessary before a signal can be transmitted; in other words, the axon has a threshold. Finally, when not conducting a signal, the potential difference across the nerve membrane, called the resting potential, is not zero, but ~ -100 mV.

CABLE THEORY

The fact that voltage signals travel along the nerve fiber immediately suggests an analysis in terms of cable theory, so called because it was originally developed to analyze the transmission characteristics of electric cables. This analysis begins with the equivalent circuit shown in Figure 7.3. Now we apply two fundamental principles:

1. Ohm's law: the potential is the current times the resistance.

2. Kirchhoff's law:
   (a) At any junction, the total current flowing toward the junction must equal the total current flowing away.
   (b) From any point in a closed-circuit loop we always return to the same potential, that is, the algebraic end sum is always zero.

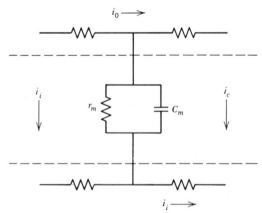

Figure 7.3 In the above circuit, C_m and r_m are the equivalent membrane capacitance and resistance, respectively.

We obtain

$$i_0 = -\frac{1}{r_0} \partial v_0 / \partial x$$

$$i_i = -\frac{1}{r_i} \partial v_i / \partial x$$

Now

$$V_m = v_0 - v_i$$

so

$$\frac{\partial V_m}{\partial x} = \frac{\partial v_0}{\partial x} - \frac{\partial v_i}{\partial x}$$

whence

$$\frac{\partial V_m}{\partial x} = -i_0 r_0 + r_i i_i$$

By Kirchhoff's law,

$$i_i + i_0 = 0$$

so

$$\frac{\partial V_m}{\partial x} = -(r_0 r_i) i_0$$

Now the change of current, i_0, with distance must just be the current through the membrane:

$$\frac{\partial i_0}{\partial x} = -i_m$$

Therefore,

$$\frac{\partial^2 V_m}{\partial x^2} = -(r_0 + r_i)\frac{\partial i_0}{\partial x} = (r_0 + r_i)i_m$$

$$= (r_0 + r_i)(i_i + i_c)$$

But from Ohm's law,

$$i_i = V_{m/r_m}$$

and

$$i_c = c_m \frac{\partial V_m}{\partial t}$$

from

$$Q = C_m V_m$$

Thus

$$\frac{\partial^2 V_m}{\partial x^2} = \frac{r_0 + r_i}{r^m} V_m + (r_0 + r_i)c_m \frac{\partial V_m}{\partial t}$$

which is

$$\frac{\partial^2 V_m}{\partial x^2} \frac{r_m}{r_0 + r_i} = V_m + r_m c_m \frac{\partial V_m}{\partial t}$$

which is the cable equation. The term $\sqrt{r_m/(r_0 + r_i)}$ is often called the space constant, λ, and $r_m c_m$, the time constant τ. Thus the cable equation may be written

$$\lambda^2 \frac{\partial^2 V_m}{\partial x^2} = V_m + \frac{\partial V_m}{\partial t}\tau$$

If we have steady-state conditions, the term $(\partial V_m/\partial t)\tau$ is equal to zero and we have

$$\lambda^2 \frac{d^2 V_m}{dx^2} = V_m$$

The solution of this is

$$V_m = Ae^{-x/\lambda} + Be^{x/\lambda}$$

For x large,

$$V_m \to 0 \therefore B = 0$$

For $x = 0$,

$$V_m = V_0 \therefore A = V_0$$

so

$$V_m = V_0 e^{-x/\lambda}$$

Therefore, for the steady-state situation, V_m will fall to $1/e$ of the value at $x = 0$ in a distance equal to the space constant. By measuring the distance required to attenuate the signal by $1/e$, we obtain λ, and from that we obtain r_m. Values of the cable parameters for several excitable membranes are given in Table 7.1.

While the cable theory supplied the rationale for determining the "circuit parameters" of the nerve fiber, it also clearly showed that the nerve fiber, however it

	λ_{cm}	τ_{msec}	$r_m k\Omega cm^2$	$c_m \mu F/cm^2$	Axon Diameter
Squid axon	0.5	0.7	0.7	1	500 μm
Crab nerve	0.25	5.0	5.0	1	30 μm
Frog muscle	0.2	24.0	4.0	6	75 μm
Nitella	2.6	21.4	21.4	1	—

Table 7.1 Cable Constants

might turn out to work, was not an ordinary electric cable. The cable equation can be written as

$$c_m = \frac{\partial V}{\partial t} = \frac{1}{r_0 + r_i} \frac{\partial^2 V}{\partial x^2} - \frac{1}{r_m}$$

Let $T = t/\tau$, where $\tau = c_m r_m$ and $X = x/\lambda$ where $\lambda = \sqrt{r_m/(r_0 + r_i)}$.

Then the equation may be written

$$\frac{\partial V}{\partial T} = \frac{\partial^2 V}{\partial X^2} - V$$

Let

$$V = u \exp(-T)$$

so

$$\frac{\partial u}{\partial T} = \frac{\partial^2 u}{\partial X^2}$$

For short pulses, $u \times \Delta t \approx 1$ and Δt is small. The solution at $x = 0$ is

$$u = \frac{1}{2\sqrt{\pi T}} \exp\left(-X^2/4T\right)$$

so

$$V = \frac{1}{2\sqrt{\pi T}} \exp\left[(-X^2/4T) - T\right]$$

Using measured values for the case of the squid axon, we find that the attenuation of a signal traveling down the axon, which goes as $\exp(-\overline{T}/\overline{X})$, is a factor of 10 for 3 cm of travel; that is, a 100-V signal falls to 3 mV when it has propagated only 15 cm along the axon. Another way of visualizing the difficulty is to realize that the resistance seen by a propagating signal in the axon is equivalent to a copper wire whose length is about 10^9 miles.

THE HUXLEY-HODGKIN APPROACH

The nerve impulse is a rapid event with a time scale of milliseconds. Because we know that generally membranes have different permeabilities for different ions, we are naturally interested in any changes in ion transport associated with the nerve impulse. However, such changes are far too rapid to be followed with any of the various methods for assaying ion concentration. This problem was surmounted by the use of the so-called voltage-clamp technique, show in Figure 7.4. In this method a potential is applied across the membrane; this potential may be restricted to some value that actually occurs during the pulse. It is assumed that if the surrounding conditions are properly maintained, the membrane will then respond as it would were that potential to occur as part of a natural signal. Thus, the membrane goes to the appropriate state for that potential and the various currents flow, only now there is a steady-state situation instead of a transient one.

Most nerve fibers are quite small and the insertion of the necessary electrodes for

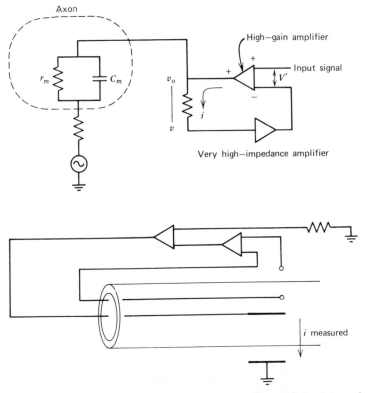

Figure 7.4 A general schematic of a voltage clamp. The parallel resistor and capacitor at left is the equivalent membrane. Shown below the schematic is a general arrangement for studying the electrical responses of an axon.

the voltage clamp, plus those for whatever measurements may be required, may be impossible owing to the small size of the fiber. However, in the 1930s, Young pointed out that the axons of the squid were very much larger and that it was an ideal experimental material. Consequently, when Huxley and Hodgkin began their classic studies on the propagation of the nerve impulse, they used voltage-clamp experiments on squid axons.

The essential results of their voltage-clamp experiments may be summarized as follows:

1. If a pulse is applied in the direction of v_{rest}, v changes rapidly and then slowly returns to v_{rest}. The membrane is not passive. From the time constant, we would expect a return to v_{rest} in about 8 μsec (see Appendix 2). In fact, the relaxation requires $\sim$ 1 to 2 msec.

3. By removing Na^+ from the bathing solution and replacing it with choline$^+$, it can be shown that the initial rise of the nerve impulse is produced by a flow of Na-ions.

The time course of events is:

1. The membrane permeability to Na^+ shows a sharp increase at a time Δt after the stimulus.

2. The permeability to K^+ increases.

3. The permeability to Na^+ decreases.

The experimentally observed ion flows and potential changes suggested an analysis based on the electrodiffusion equation. The most useful form of this equation was derived by Goldman, who began by assuming that the electric field across the membrane is constant. In that case, the potential gradient can be written as the potential difference across the membrane divided by the membrane thickness. If we assume that the membrane is in hydrostatic equilibrium,

$$dP/dx = 0$$

and then we have

$$J = -\omega cRT \left(\frac{1}{c} \frac{dc}{dx} + \frac{\bar{v}}{RT} \frac{dP}{dx} + \frac{zF}{RT} \frac{d\psi}{dx} \right)$$

$$= -\omega RT \left(\frac{1}{c} \frac{dc}{dx} + \frac{zF}{RT} \frac{\partial \psi}{\partial x} \right)$$

or

$$-J \times \frac{1}{\omega RT} = \frac{dc}{dx} + \frac{czF}{RT} \frac{\partial \psi}{\partial x}$$

which has the solution:

$$c = A \, \exp\left(-\frac{zF}{RT}\frac{\Delta\psi}{\Delta x} x\right) - \frac{J}{wRT} \frac{RT}{zF} \frac{\Delta x}{\Delta \psi}$$

At $x = 0$, $c = c_1$, and at $x = \Delta x$, $c = c_2$, which yields

$$-J = \frac{zF\Delta\psi}{RT\,\Delta x} \omega RT \left[\frac{c_2 - c_1 \exp\left(\dfrac{zF\Delta\psi}{RT\Delta x}\Delta x\right)}{1 - \exp\left(-\dfrac{zF\Delta\psi}{RT\Delta x}\Delta x\right)} \right]$$

Since in this particular case we measure currents and not fluxes, we want to change J to I. First, we transform the molar mobility ω to the electrical mobility u by

$$u = z\omega F$$

For our case, since we are interested in Na^+ and K^+, $z = +1$ and

$$I_+ = -\frac{u_+ F\Delta\psi}{\Delta x} \left[\frac{c_2 - c_1 \exp(F\Delta\psi/RT)}{1 - \exp(F\Delta\psi/RT)} \right]$$

Now

$$\frac{a - be^x}{1 - e^x} = -\frac{ae^{-x} - b}{1 - e^{-x}} \qquad \text{by} \qquad LHS \times \left(\frac{e^{-x}}{e^{-x}}\right)$$

so

$$I_- = \frac{u_- F\Delta\psi}{\Delta x} \left(\frac{c_z e^{-F\Delta\psi/Rt} - c}{1 - e^{-F\Delta\psi/RT}} \right)$$

If no external current is passing, electroneutrality must apply, and therefore $I_+ + I_- = 0$, whence

$$(u_+ c_1 + u_- c_2) \exp(-F\Delta\psi/RT) = u_+ c_2 + u_- c_1$$

or

$$\Delta\psi = \frac{-RT}{F} \ln \frac{u_+ c_2 + u_- c_1}{u_+ c_1 + u_- c_2}$$

which is Goldman's equation.

Huxley-Hodgkin assumed that membrane current was given by

$$I = I_{K^+} + I_{Na^+} + I_{Cl^-}$$

whence Goldman's equation becomes

$$\Delta\psi = -\frac{RT}{F} \ln \frac{[K_2] + [Na_2] + [Cl_1]}{[K_1] + [Na_1] + [Cl_2]}$$

The concentrations can also be written as permeabilities. Now

$$I_K = \frac{u_K F \Delta\psi}{\Delta x} \frac{(C_K)_0 - (C_K)_a e^{-F\Delta\psi/RT\Delta x}}{1 - e^{-\Delta\psi F/\Delta x RT}}$$

Now, since

$$(C_K)_0 \sim [K_0]$$

in the bathing solution,

$$(C_K)_0 = \beta_k [K_0]$$

and similarly

$$(C_K)_a = \beta_K [K_i]$$

Hence if we define P_K, the permeability coefficient, as

$$u_K \beta_K \frac{RT}{\Delta x F}$$

then

$$I_K = P_K \frac{F^2 \Delta \psi}{RT} \frac{(K)_0 - (K_i) e^{-\Delta \psi F/RT}}{1 - e^{-\Delta \psi F/RT}}$$

and similarly for I_{Na}, I_{Cl}.
Then we can write

$$w = (K_0) + \frac{P_{Na}}{P_K} (Na)_0 + \frac{P_{Cl}}{P_K} (Cl)_i$$

$$y = (K_i) + \frac{P_{Na}}{P_K} (Na)_i + \frac{P_{Cl}}{P_K} (Cl)_0$$

and the total current is

$$K = \frac{F^2 \Delta \psi}{RT} P_K \frac{w - ye^{-\Delta \psi F/RT}}{1 - e^{-\Delta \psi F/RT}}$$

The potential across a membrane when no ionic current flows is E, that is, $V = E$ when $I = 0$, so

$$E = \frac{RT}{F} \log_e \frac{y}{w}$$

and the membrane conductance, g, is

$$g = (dI/dV)_{I \to 0}$$
$$= \frac{F^2 P_K}{RT} \left[V \frac{d}{dx} \left(\frac{w - ye^{-VF/RT}}{1 - e^{-VF/RT}} \right) + \left(\frac{w - ye^{-VF/RT}}{1 - e^{-VF/RT}} \right) \right]$$

$$g = \frac{F^3}{(RT)^2} \, EP_K \left(\frac{wy}{y - w} \right)$$

that is, we get P_K if we know P_{Na}/P_K and P_{Cl}/P_K.

The above analysis permits one to interpret the voltage-clamp data. Huxley and Hodgkin developed a formalism to describe the results of their voltage-clamp experiments in an analytical way. They argued as follows. Let us consider the conductance, g, of the membrane for a particular ion; the conductance is a reciprocal resistance, is measured in mhos/cm^2 of membrane surface, and, in the case of the excitable membrane, will be a function of time. Hence, we have g_k, g_{Na}, and g_L for, respectively, potassium, sodium, and chlorine.

All potentials are measured with respect to the resting potential. Then

$$V_{Na} = E_{Na} - E_r$$
$$V_K = E_K - E_r$$
$$V_L = E_L - E_r$$
$$V = E - E_R$$

The current density is then

$$J = C_m \frac{dv}{dt} + J_i$$

that is, we divide the membrane current into a capacitative part and an ionic part, where

$$J_i = \Sigma \, J_{Na,\,K,\,L}$$

Then

$$J_{Na,\,K,\,L} = g_{Na,\,K,\,L} \, (V - V_{Na,\,K,\,L})$$

where $V_{Na,K,L}$ are the equilibrium potentials.

Hodgkin and Huxley assumed the ion flows could be described in terms of channels opening and closing; the data suggested that potassium and sodium flows required, respectively, four and three "molecular events" for the formation of a channel. Thus, if n is the probability of one of the events, then

$$g_K = g_K^{max} n^4$$

The parameter is assumed to obey a first-order rate equation:

$$\frac{dn}{dt} = \alpha_n(1 - n) - \beta_n n$$

where α_n and β_n are the rate constants and, under fixed conditions, are functions only of the membrane potential. As the inside of the axon becomes more positive, α_n increases and β_n decreases. A similar result holds for sodium: three events, each of probability m, create a channel, while one event, of probability $(1 - h)$ closes it. Thus we have the probability that the channel is opened is $m^3 h$ and we write

$$g_{Na} = g_{Na}^{max} m^3 h$$

These two parameters are also assumed to follow a first-order equation:

$$\frac{dm}{dt} = \alpha_m(1 - m) - \beta_m m$$

and

$$\frac{dh}{dt} = \alpha_h(1 - h) - \beta_h h$$

where, as before, α and β are functions of the membrane potential under a set of fixed experimental conditions.

The rate constants are given in Table 7.2. They were determined at $T = 6°$ and increase by about a factor of three for a temperature increase of $10°$.

Table 7.2 Rate Constants

α_n	$= (0.01)(V + 10)/[e^{(V+10)/10} - 1]$
β_n	$= 0.125e^{V/80}$
α_m	$= (0.1)(V + 25)[e^{(V+25)/10} - 1]$
β_m	$= 4e^{V/18}$
α_h	$= 0.07e^{V/20}$
β_h	$= [e^{(V+30)/10} + 1]^{-1}$

Now we can write an expression for the total membrane current as a function of time and voltage:

$$I = c_m \frac{dV}{dt} + g_K n^4 (V - V_K) + g_{Na} m^3 h(V - V_{Na}) + g_L (V - V_L)$$

where the potentials are given in mV, the currents in $\mu A/cm^2$, the conductances in mmho/cm^2, the capacitance in $\mu F/cm^2$, and the time in milliseconds. Typical values of some of these quantities are:

1. For potentials: Na ~ -115; K $\sim +12$; and Cl ~ -10.6 mV.

2. For conductances: Na ~ 120; K ~ 36; and Cl ~ 0.3 mmho/cm^2.

Of course, these are functions of time.

Consider the schematic axon shown in Figure 7.5. By Ohm's law,

$$r_1 I_1 = -\partial V_1 / \partial x$$

and

$$r_2 I_2 = -\partial V_2 / \partial x$$

Now

$$I = \partial I_1 / \partial x = -\partial I_2 / \partial x$$

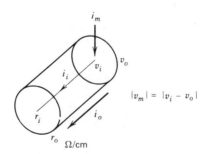

$$|v_m| = |v_i - v_o|$$

Figure 7.5 The electrical quantities that can describe an axon.

and

$$V = V_2 - V_1$$

Therefore,

$$I = \frac{1}{r_1 + r_2} \frac{\partial^2 V}{\partial x^2}$$

where I is the membrane current per unit length.

Since the axon is surrounded by a large volume of conducting fluid,

$$r_1 \ll r_2$$

and

$$I \doteq 1/r_2 (\partial^2 V / \partial x^2)$$

Let us transform to the current density, J amp/cm^2, and convert r_2 to the specific resistance or resistivity, R_2, of the axoplasm. Then

$$J = \frac{a}{2R_2} \frac{\partial^2 V}{\partial x^2}$$

Whence we obtain

$$\frac{a}{2R_2} \frac{\partial^2 V}{\partial x^2} = c_m \frac{\partial V}{\partial t} + g_K n^4 (V - V_K) + g_{Na} m^3 h (V - V_{Na}) + g_1 (V - V_L)$$

Let θ be the velocity of the impulse. The shape of the nerve impulse remains constant as the impulse travels down the axon. Therefore, the solution to the above equation must also be the solution to a wave equation. However, we know that the wave equation will have the form:

$$\frac{\partial^2 V}{\partial x^2} = \frac{1}{\theta^2} \frac{\partial^2 V}{\partial t^2}$$

$$\frac{a}{2R_2} \frac{\partial^2 V}{\partial x^2} = \frac{a}{2R_2\theta^2} \frac{\partial^2 V}{\partial t^2}$$

and this can be substituted in the previous result. That equation must be solved numerically; the easiest way is by iteration on a computer. Begin by guessing a value of θ and iterate. The value of θ that is correct will be the one for which the potential returns to the resting value. Of course, θ can also be measured directly in the laboratory, and so the theory can be tested. The ability of the above analysis to give an accurate representation of the nerve signal can be seen from inspection of Figure 7.6.

Hodgkin and Huxley's analysis has been supported by a variety of experimental studies. Their fundamental conclusion, that the nerve impulse is a transient reversal of the potential across the membrane produced by a rapid change in the permeability to Na^+ and then to K^+, is now beyond doubt. What is the detailed behavior of these ionic currents? First, the nonzero resting potential is the result of a concentration gradient across the membrane produced by the active transport of K^+-ions. Thus once again, we see active transport as a fundamental membrane property. When a stimulus is applied, the membrane potential is less negative; such a step is said to depolarize the membrane. If we depolarize the membrane with a voltage clamp, the current first flows inward; then there is a rapid reversal to an outward flow. The membrane current in this case has two components: the inward component is due to Na^+ and the outward component to K^+. The observed behavior occurs because the permeability to Na increases and then decreases after a short time, while the K permeability does not increase immediately, but does remain roughly constant once it has increased, until the depolarization is removed.

More recent voltage-clamp studies show that there are complications. For example, if there is no external sodium, there can be no initial Na-current and therefore presumably no action impulse. In fact, there are systems, such as snail axons, in which Ca^{++} can substitute for Na^+. These systems also show a combination of the slow outward K-flow with a much faster outward K-current. Outward K-currents are not easily studied, but recent work by Neher and Lux has overcome many of the difficulties. They used an ion-selective microelectrode to measure the K-current outside the neuron. They find that the total K-flow from the cell matches the outward current from the nerve if one assumes that the current is due to K.

More recent work shows that it is possible to make the above analysis somewhat more economical. However, what is now required is a detailed molecular mechanism from which the above general representation could be made to appear naturally.

Progress in that direction may be possible by considering not only the nerve signal but also the noise associated with it. As you probably know, no electrical signal is "pure," but is always accompanied by voltage variations of a random character. We usually describe this situation by giving the signal to noise ratio, S/N. A number of physical processes are involved in the production of noise; in the case of the nerve membrane, one such process is the random thermal motion of charge carriers, which produces Nyquist noise. Flicker noise should also be observed for very small currents, and a random resistance fluctuation should be present as the various "channels" through the membrane are opened or blocked. This latter noise is very difficult to observe because the flicker noise is so much larger. However, Stevens has shown and

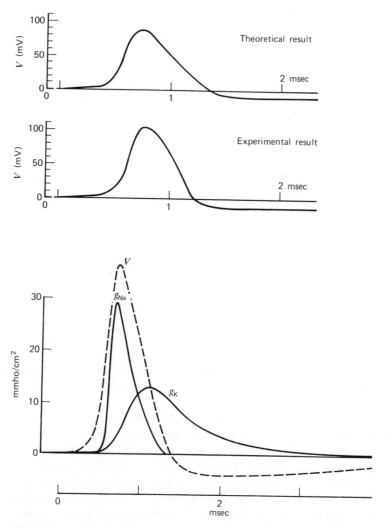

Figure 7.6 The two upper curves show how well the Huxley—Hodgkin solution actually predicts the observed action potential. The third graph shows how the conductances change as the action potential develops.

Fishman has observed that the conduction noise is flat for frequencies ~ 10 to 10^2 Hz and falls off as $1/f^2$ at higher frequencies. Furthermore, if the transport of potassium is blocked, the noise goes to zero.

ELECTRICAL ACTIVITY OF THE CENTRAL NERVOUS SYSTEM

The constant flow of electrical signals along the nerve fibers naturally suggest that some electrical activity might be observable in the brain. It is not obvious that this would be of any interest, since the sum of numerous random voltage spikes might simply produce a constant dc potential. In fact, it turns out that the brain does show rather remarkable electrical activity, whose origin and interpretation remains for the most part obscure.

Part of this obscurity is probably self-induced by the overly simple emphasis of this chapter. We have focused on the conduction process in the axon of the nerve cell because it is this process that is at the same time dramatic, not too hard to observe, and susceptible to analysis. However, the axon is but one part of a neuron, and it is certainly a mistake to see the neuron as involved only with signals moving along the axon. We must replace the notion of a neuron with relatively passive dendrites by a model in which the neuron is simultaneously involved in many different pathways by the interaction of the dendrites. This leads unavoidably to the idea the neuron interaction can occur through these dendrites. Here, however, the signals are not in the form of spike potentials but occur as small changes in the membrane potential. The apparent implication is that if the neurons are close together, close enough for the dendrites to connect, then the attenuation of the signal in these short fibers is not very important. This means that it is not very important to generate spikes, which can be thought of now as the way long-distance signals are rapidly propagated.

Considerations of this sort make it clear that we now need to begin to recognize the complexity of neuron organization and neuron processes. For example, the Nissl bodies are the site of extensive protein synthesis; roughly one-third of the protein in the cell body is synthesized per day. This protein is transported down the axon or out of the cell. Of course, the axon cannot be the site of any protein synthesis because it does not contain ribosomes.

The nervous system has two major parts. First is the central nervous system, (CNS), composed of the brain, brain stem, and spinal cord. The second component is the peripheral or autonomic nervous system (ANS), which controls involuntary responses such as heart action. The autonomic nervous system is itself divisible into two components, the sympathetic nerves that are involved with physiological "fight or flight" responses, such as adrenal gland activation, and the parasympathetic system that is involved with relaxation responses. The distinction between the CNS and ANS is seen in the properties of the neurons that compose them. For example, the number of neurons in the CNS are fixed from birth at about 10^{12}; they increase in size and connectiveness, but not number, and if a CNS neuron is injured or destroyed it is not repaired or replaced. In contrast, ANS neurons can be both repaired and replaced by regeneration from the body cell of the neuron. The neurons of the two parts differ in other ways. For example, CNS neurons are much more sensitive to oxygen deprivation than ANS neurons; CNS neurons begin to die after about 4 to 5 minutes without oxygen, whereas ANS neurons can survive for 2 to 3 hours.

One other example of specialization will be given. Axons can be coated with a sheath of lipid; such nerves are said to be myelinated and they appear white and shiny. In other cases, the axon is nonmyelinated and the nerve appears gray in color. This is the origin of the gray and white matter of the nervous system, seen for example in the outer (white) region of the spinal cord versus the inner (gray) and in the outer and inner regions of the brain.

The electrical activity of the brain is not too difficult to observe and can be recorded by simply attaching electrodes directly to the scalp. The recording, called an electroencephalogram or EEG, must be made with the proper electrical filtering in order to avoid confusing the record of the brain's electrical signals with signals produced by such things as muscular movement.

The EEG shows two general components:

1. A continuous and essentially periodic variation called the spontaneous activity.

2. Localized signals, whose amplitudes are larger than those of the spontaneous activity. These localized signals, called evoked potentials, certainly originate from the input of the sensory receptors.

The spontaneous activity is usually marked by the presence of a dominant frequency, which can be used to characterize the EEG pattern: If the frequency is ~ 8 to 14 Hz,

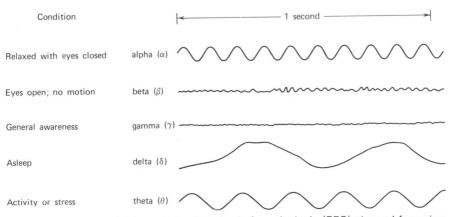

Figure 7.7 The general forms of electrical signals from the brain (EEG) observed for various conditions of the subject.

the activity is called an α-wave; if 4 to 8 Hz, a θ-wave; if 14 to 60 Hz, a β-wave; and if < 4 Hz, a δ-wave. In general, the frequency of waves recorded from the frontal and parietal regions are higher than from the occipital; it is also found that the α-wave, dominant in the waking state of a "relaxed" adult brain, is suppressed by visual stimuli. The α-dominant pattern does not develop much before the age of 12; during childhood, the θ-pattern is usually dominant. Brain activity in infants is bimodal: a δ-pattern mixed with a fairly high-frequency rhythm of 40 ± 10 Hz.

Clearly, all of this activity is somehow connected with electrical activity in brain cells of the cortex. The space constant, λ, of such a cell is large, and therefore the dipolar aspect of such a cell is insignificant. This is not the case for the fine dendrites; in that case λ is less than the length of the dendrite. Therefore, an electrode on the scalp will "see" the dendrites as dipoles. From that, one may conclude that the dendritic activity, at least from the observable areas of the upper third of the cortex, is nonrandom, as Figure 7.7 shows.

chap. 8

Artificial Membranes

The complicated behavior exhibited by those membranes produced by biological processes has served to stimulate considerable efforts aimed at the production of artificial membranes of known composition that can be studied under controlled conditions. This effort takes its cue from Langmuir's work on the production of monolayers formed by spreading on a clean liquid surface. Although Langmuir did not try to extend his experiments on monolayers to the production of artificial membranes of biological significance, his experiments contained the essential techniques for such an effort.

PREPARATION AND PROPERTIES

Nevertheless, a layer spread on the surface of a liquid is somewhat difficult to work with; furthermore, it is a monolayer, while we wish to at least approximate the biological membrane, which is bilayer. Simple bilayers can be made by taking advantage of the natural flow of a lipid that has been suspended on a small aperature. This can be done in the following way: the side of a thin-walled teflon cup can be further thinned by compressing the wall with the heated tips of the jaws of a pair of pliers. Then this region can be punctured with a hot needle. The cup is placed inside a large beaker; this beaker is filled with an electrolyte, as is the cup. A small drop of lipid is then "painted" onto the hole with a very fine artists' camel's brush or placed with a microsyringe. This drop will spontaneously thin and the center of the drop should, if all goes well, become a bilayer lipid membrane (BLM) in about 15 minutes or less. Of course, many difficulties, such as vibration, can lead to the breakage of the fragile membrane. The long lipid molecules are almost certainly arranged in the BLM so that they are perpendicular to the membrane surface, the interface between the membrane and the solution. One argument in favor of this is the general result that each fatty acid molecule in a monolayer always appears to occupy about 22×10^{-16} cm^2 of the monolayer surface, independent of the length of the fatty acid chain. Furthermore, as you already know, the lipid molecule is often hydrophobic at one end and hydrophyllic at the other. The lowest energy state will be the one in which the molecules of the lipid are perpendicular to the membrane sheet, with the hydrophobic end facing away from the aqueous phase. This is borne out by the following argument. We consider the direction for a particular end, and ask, is the end pointed toward the lipid or toward the solution? We can write the equilibrium constant, K, as just the concentration ratio for the two cases. Thus

$$[N_0]/[N_{sol}] = K$$

Now

$$[N_0] = A \exp(-E_0/kT)$$

and

$$[N_{sol}] = A \exp(E_{sol}/kT)$$

so

$$K = \exp[-(E_0 - E_{sol})/RT] = \exp(-\Delta F/RT)$$

where ΔF is the free-energy difference between the two states and is ~ 1.7 kcal/mole. Thus

$$K \doteq 5 \times 10^{-12}$$

and the chance that a lipid molecule is "backward" is very small indeed. Finally, we note that optical birefringence studies also show that the lipids are perpendicular to the membrane surface, both in artificial and biological membranes.

Since the biological membrane is, among other things, a barrier, and particularly a barrier that separates two different regions, an obvious topic for investigation with artificial membranes is the surface physics and interfacial phenomena associated with them.

The most obvious interfacial phenomena is surface tension; in fact, Langmuir's early efforts were aimed at measurements of this quantity for different substances. Suppose we consider a liquid in contact only with its own vapor. If we wish to increase the area of this interface, we must add molecules and these must come from the body of the liquid below the surface. Since there are cohesive forces acting between the molecules of the liquid, work must be done to move a molecule from the body of the liquid to the surface. Therefore, the molar-free energy of the surface is greater than the molar-free energy of the interior of the liquid, and a molecule at the surface has a potential energy greater than the potential energy of a molecule in the body of the fluid. Thus, the surface potential energy of a liquid will increase with the surface area. Since the minimum energy is the most favorable state, the surface tends to a shape with the smallest surface area. For this reason, a water drop tends to be spherical, which is the smallest surface area for a given volume.

It has been known for at least 150 years that the mechanical behavior of a liquid surface could be described in terms of an infinitesimally thin imaginary membrane that covers the surface and exerts a force, parallel to the surface, which opposes any increase in the surface area. This force, taken per unit length, is known as the surface tension. It can be measured directly in a Langmuir balance, shown schematically in Figure 8.1; in this apparatus, the force tending to minimize the surface area is balanced by the force $\bar{F}$ exerted in the opposite direction. Suppose the width of the film is ℓ. Then the work done by actually moving the bar against the surface force is $F \times \Delta x$,

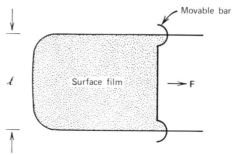

Figure 8.1 The Langmuir balance is used to determine the surface tension of thin layers, particularly monolayers, of fluids.

which must be the same as the surface potential energy/cm^2. The work done increases the area of the film by $2\,l\,\Delta x$, so the surface tension σ is

$$\sigma = \frac{F\Delta x}{2\,l\,\Delta x} = \frac{F}{2l}$$

From the fact that σ is constant for films of varying thickness, we conclude that the surface energy really is essentially localized in a very thin surface membrane.

The above argument can now be extended so that we can understand the behavior of the lipid drop when it is painted on the aperture of the Teflon cup. Let us introduce the interface potential energy/cm^2, Γ. We must consider the force between the molecules of the liquid and those of the solid, as well as the force between the molecules of the liquid. These two forces are often called adhesive and cohesive forces.

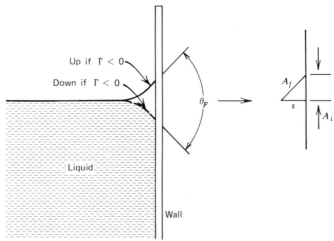

Figure 8.2 This drawing illustrates the quantities used in analyzing surface tension.

Now if $\Gamma > 0$, the liquid cannot adhere or "wet" the surface. We recognize the following possibilities: $\Gamma < \sigma$, $\Gamma = \sigma$, and $\Gamma > \sigma$, which correspond, respectively, to some liquid-solid attraction, no attraction, or actual repulsion. On the other hand, if $\Gamma < 0$, wetting will occur, while if $\Gamma = 0$, the forces will exactly balance.

The angle of contact between the liquid and solid can be calculated. In Figure 8.2, let the area of the interface be A_i and the area of the fluid side, A_f. The total potential energy is

$$U = \sigma A_f + \Gamma A_i$$
$$= \frac{\sigma s}{\sin \theta} + \frac{\Gamma s}{\tan \theta}$$

Then

$$dU/d\theta = \frac{-\sigma s \cos \theta}{\sin^2 \theta} - \frac{\Gamma s}{\sin^2 \theta}$$

Setting this equal to zero yields

$$\cos \theta = \frac{-\Gamma}{\sigma}$$

Thus when the droplet is placed on the aperture, the Teflon is not "wettable." The edges of the droplet "turn up" around the aperture, and lipid flows from the center toward the ring. The lipid thins in the center and thickens around the edge. The thickened annular ring is directly observable and is called the Plateau-Gibbs border. At some point equilibrium must occur. The forces acting are the mechanical "suction" involved in the flow to form the Plateau-Gibbs border and the van der Waals attraction; these two are opposed by electrical forces in the double layer and probably other contributions. In many cases, the equilibrium occurs when the central region of the original drop has thinned to lipid bilayer.

The surface tension of the BLM may be determined directly through the application of Laplace's equation:

$$P = \frac{1}{\sigma} \left(\frac{1}{R_1} + \frac{1}{R_2} \right)$$

Suppose we have a slight pressure difference across the BLM; such a difference could be produced by adding a very small amount of fluid to the inner cup. A pressure increase may be produced that will change the radius of curvature of the bulge by dR.

The pressure increase does work against the surface tension of the membrane:

$$dw = P \times \text{area of bulge} \times dR$$

So

$$dw = P\frac{4\pi R^2}{2}\, dR = 2\pi PR^2\, dR$$

The work required to expand the bubble is, taking it to an ellipsoidal surface,

$$dw = 16\pi R\, dR\, \sigma$$

whence

$$P = \frac{8\sigma}{2R}$$

Thus, by measuring the radius of the bulge produced by a small-pressure difference ($\sim 10^{-2}$ cm H_2O), we can find the surface tension of the membrane, σ.

Values of σ for the simplest oil-water systems are as large as 35 dynes/cm; typical values of σ for "ordinary" BLMs are ~ 2 dynes/cm. How do these values compare with the surface tension of a biological membrane? The first measurements of that quantity were carried out by Harvey, who used a centrifugal technique in order to measure directly the force parallel to the membrane in sea urchin eggs. By carefully increasing the effective g, he was able to show that cells were pulled apart when the effective force exceeded about 0.2 dynes/cm. Later measurements seem to indicate that surface tensions as low as 0.04 dynes/cm may occur. These values are considerably less than those observed for pure BLM. However, the values are not directly comparable because the biological membrane also contains a protein coat. The incorporation of protein into BLMs in a structurally significant way is not a simple matter; however, determinations of σ for such systems do yield values considerably less than those found for pure BLM. Hence, the small values of σ observed for biological membranes presumably are due to the incorporated proteins in those membranes.

The surface tension problem should not be used as an argument against the view that BLM systems are fundamentally good models of biological membranes. A really eseential requirement for a model system is that it be the same thickness as the "real" membrane, and BLMs are very satisfactory in this regard. Optical methods, capacitance methods of the type pioneered by Frike, and direct examination by electron microscopy all confirm that the thickness of the BLM is essentially the same as the thickness of biological membranes. Thus presumably, we can duplicate the properties of biological membranes by the addition of the proper protein complexes, provided we can identify, obtain, and learn how to incorporate them in the BLM.

THICKNESS DETERMINATION

Let us consider how the thickness of a BLM can be determined optically; this approach makes use of observation of the fringes produced b interference between the light waves reflected from the front and rear surfaces of the BLM. The path difference between two rays is

$$\Delta = 2n_{BLM}t \cos \theta$$

where n_{BLM} is the index of refraction, θ is the angle of refraction, and t the thickness of the BLM. The phase difference is then

$$\delta = \pi 2n_{BLM}t \cos \theta / \lambda$$

For normally incident light, the reflected intensity is calculated with the aid of the Fresnel equation:

$$E = \frac{r_1 + r_2 \exp(-2i\delta)}{1 + r_1 r_2 \exp(-2i\delta)}$$

where

$$E = \frac{\left(\dfrac{n_s - n_{BLM}}{n_s + n_{BLM}}\right) + \left(\dfrac{n_{BLM} - n_s}{n_{BLM} + n_s}\right) \exp(-2i\delta)}{1 + \left[\dfrac{n_s^2 + 2n_{BLM}n_s - n_{BLM}^2}{(n_{BLM} + n_s)^2}\right] \exp(-2i\delta)}$$

and n_s is the index of refraction of the solvent.
Since

$$\frac{n_s - n_{BLM}}{n_s + n_{BLM}} = -\frac{n_{BLM} - n_s}{n_{BLM} + n_s}$$

and the BLM is usually in an aqueous solution, we can write the ratio of the reflected to incident light as

$$\frac{I}{I_0} \cong 4 \left(\frac{n_s - n_{BLM}}{n_s + n_{BLM}} \right)^2 \sin^2 \delta$$

A properly prepared lipid drop will thin; eventually it ceases to exhibit interference fringes. For an ordinary BLM,

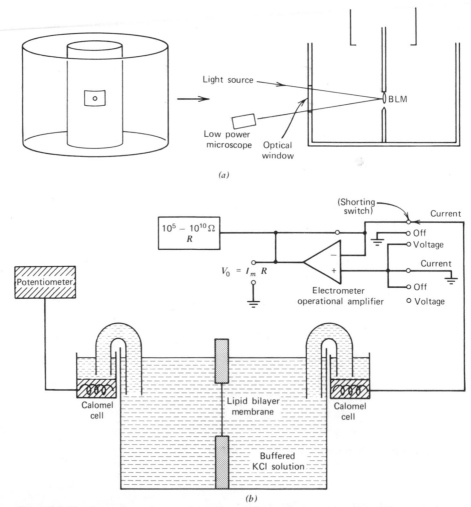

(a)

(b)

Figure 8.3 This figure shows the experimental setup for producing and studying bilayer membranes and for measuring their electrical properties.

$$\frac{I}{I_0} \cong 10^{-3} - 10^{-4}$$

and thus the BLM becomes black, or essentially nonreflecting, at least to the eye.

The remarkable ion transport properties of biological membranes are not yet reproducible by BLM systems; in spite of obvious expected differences, the electrical properties of BLMs have proved to be very interesting. Measurements of the electrical conductivity of BLMs are surprisingly difficult, and individual determinations are subject to considerable variation. These difficulties apparently result from the variations of solvent inclusions in the BLM. The resistance ranges from 10^4 to 10^{10} Ω/cm^2, although the addition of a number of substances will produce obvious and even large changes in the resistance. Of course, the actual mechanism of conduction remains obscure. The situation with respect to the capacitances of BLMs is quite different. In this case, the determinations are quite repeatable. If the BLM is in a stable state, we would expect the capacitance to depend only on the geometry. We observe that C is independent of the applied frequency as long as the applied dc potential is less than about 50 mV. Values of C range from 0.3 to 1.3 $\mu F/cm^2$. Experimental methods for carrying out these measurements are shown schematically in Figure 8.3.

chap. 9

Energy Trans-
duction Proc-
esses

Energy transduction is the process by which energy is changed from one form to another; the mechanism that accomplishes the change is the transducer. Although this discussion is mainly concerned with processes of transduction at the cellular level, the term need not be restricted to cellular contexts; for example, the heart muscle is a chemicomechanical energy transducer.

How is the energy required by cells supplied? At the most fundamental level, that energy is supplied by the sunlight, which drives photosynthesis in the case of plants, and by the ingesting and breaking down of complex organic molecules, mostly by oxidation reactions, in the case of animals. Different bacteria not only can carry out reactions of the above types but also can obtain energy from simple inorganic reactions.

The above distinctions are not absolute. For example, when light is not available, plants live on energy whose origin is the oxidation of organic molecules built up during the photosynthetic period. It is also correct to see photosynthesis as the most fundamental process for animal life, because of the metabolic limits of animal cells. Although animals cells can make, given a correct input of all the necessary sugars, most fatty acids, all sterols except vitamin D (vitamins are made by plants and microbes, but not animals), all purines and all pyrimidines, they cannot make 10 essential amino acids; of course, plants can.

Thus the initial process, for say a mammal, is to get some very large organic molecules — proteins, polysaccharides, and neutral fats — into the alimentary canal where they can be broken down to amino acids, monosaccharides, fatty acids, and glyerol and absorbed into the circulation

Note that all of the free energy produced in this breakdown goes into heat. It is the change in ΔF that drives the metabolism; overall, from start to product, ΔF is negative and usually large. A large ΔF guarantees that the reaction will proceed and that the products will be $\gg$ the reactants. However, it says nothing about the speed of the reactions and this of course is where the enzymes come in.

Once the products of the breakdown have been brought into circulation, they can begin to be processed to make energy available. This energy is required for processes ranging from moving ions and molecules in active transport to contracting muscle.

Energy is also stored in the form of insoluble compounds like starch or glycogen, which must be processed before the energy becomes available. In higher animals and plants, the first steps transform the stores into soluble sugars. Then these are distributed to the cells. At the cellular level, the sugar is oxidized to produce H_2O, CO_2, and energy.

The processes carried out by living matter are seldom the direct consumers of the energy that is immediately available to the cell. Instead, these processes are driven by energy derived from an intermediate source, which is the hydrolysis of ATP:

$$\text{ATP} + H_2O \rightleftharpoons \text{ADP} + H_3PO_4 + \Delta G$$

$$\Delta G \doteq -7 \text{ kcal/mole}$$

Because of the central role of ATP, the problem of energy transduction has been dominated by the study of the processes by which incoming energy may be used to produce ATP.

Of course, the hydrolysis of ATP does not simply release energy. That would not lead to useful work but only to the generation of heat. Instead, the hydrolysis reaction is coupled in some way to the cellular process that requires the energy. For example, since the formation of a bond in the construction of a protein requires a free energy $\Delta G \doteq \frac{1}{2}$ kcal/mole, the free energy of hydrolysis of ATP will insure that the equilibrium of the process is shifted strongly toward the formation of the bond. Although we may represent the process by which the cell makes use of the ATP hydrolysis energy by some schematic equation such as

$$ATP + X \rightarrow ADP + X \sim P + \Delta G$$

where X is some compound activated by the ATP, in fact, we have no idea of how such energy transfers actually occur. This problem is one of the most challenging in biophysics.

ATP is produced by different processes in different biological systems. To date, we recognize four distinct processes:

1. Oxidation-reduction reactions that occur on the inner membranes of mitochondria in connection with cellular respiration.

2. The photosynthetic processes in green plants, which occur on the membranes of the grana in chloroplasts.

3. Reactions that occur on the inner membranes of bacteria in conjunction with bacterial metabolism.

4. Light driven reactions localized in the chromatophores of photosynthetic bacteria.

It is certainly significant that all of these processes are associated with a membrane. In addition, we can split the above processes into two groups. One group depends on respiration, that is, on having oxygen available; the other group is photosynthetic, that is, driven by light. A schematic representation of these two general processes is shown in Figure 9.1. We should also note that membranes are the mediators between chemical energy, in the form of ATP, and electrical energy in the form of the membrane potential and the ion flows.

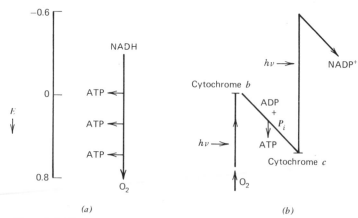

Figure 9.1 Electron transport in the respiratory chain of mitochondria and in photosynthesis. (a) Mitochondrial process. (b) Photosynthesis.

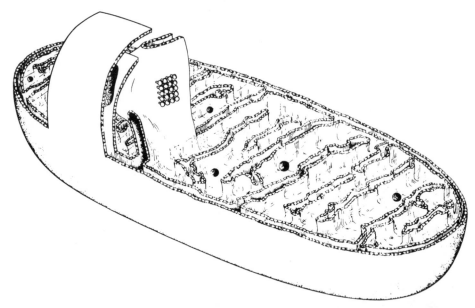

Figure 9.2 Mitochondria are generally sausage-like organelles that average about 3 μm long and about 0.5 μm in diameter. They were one of the first organelles to be studied in detail, thanks to the development of centrifugation techniques permitting them to be isolated in large quantities from homogenized liver. If this is spun at 1500 g for about 3 minutes, the large fragments and the nuclei are deposited as a pellet. The remainder of the suspension is then spun at 2000 g for one-half hour; the pellet so formed contains largely mitochondria. Biochemical studies clearly show the ability of the organelles to carry out *in vitro* oxidation of fatty acids to H_2O and CO_2. The outer membranes of the mitochondrian control the flow of inorganic ions and organic molecules, as well as the release of the ATP required by the cell processes. The complexly folded inner membrane is the site of the oxidative phosphorylation that produces ATP.

MITOCHONDRIAL PROCESS

Let us consider in more detail the first of the four processes given above — the production of ATP by mitochondria — whose structure is shown in Figure 9.2. The mechanism of this process is the subject of three different hypotheses, each of which has implications for the other process in our list. These hypotheses are:

1. A purely biochemical view, the so-called chemical hypothesis, maintains that the essential event is the formation of a short-lived, high-energy intermediate through a series of purely chemical reactions.

2. The chemiosmotic hypotheses of Mitchell, in which it is argued that the essential event is the translocation of H^+-ions across the membrane without either a cation flow in the opposite direction or an anion flow in the same direction. This produces a membrane potential that can then drive ATP synthesis.

3. The conformational change hypothesis of Green in which it is argued that the energy required to do the work of folding protein structures, and which is available upon relaxation, is connected to ATP synthesis.

Various items of experimental evidence can be advanced for each of these views.

For the conformational hypothesis:

1. It is known that ATP alters the conformation of myosin; this is the essential step in the process of muscle contraction. The reverse process also appears to occur: ATP can be synthesized through the elongation of a contracted fiber that has been chemically prepared. Thus, there is at least one case in which ATP synthesis is linked to mechanical change.

2. Gross changes in the morphology of the mitochondrial membranes occur in response to changes in the energy state of the mitochondrian. Critics of this hypothesis question whether the time scale of these changes is appropriate to the time scale of ATP formation, which does appear to proceed more rapidly.

For the chemiosmotic hypothesis:

1. In the case of the red blood cell membrane, a reversal of the normal flows of Na^+ and K^+ across the membrane leads to the synthesis of ATP.

For the chemical hypothesis:

1. Spectroscopic evidence strongly suggests that certain cytochromes, compounds known to be localized in the membrane, do have high-energy forms.

CHLOROPLAST PROCESS

The complexities and uncertainties of the mitochondrial ATP problem extend to the second process in our list, photosynthesis. Indeed, in a preliminary remark to a discussion of some recent work in this area, the editor of *Nature* observed that "reading the literature of photosynthesis is almost certainly bad for the health." However, we can see that both processes share a significant feature, in addition to the membrane link. The simplest view of the mitochondrial result is that

$$\text{Glucose} + \text{oxygen} \rightarrow \text{carbon dioxide} + \text{water} + \text{energy}$$

or

$$C_6H_{12}O_6 + 6O_2 \rightarrow 6CO_2 + 6H_2O + \Delta H = 672 \text{ kcal/mole}$$

This really is not quite right and in order to see where the ATP comes from, we should write:

$$C_6H_{12}O_6 + 6H_2O + 6O_2 \rightarrow 6CO_2 + 12H_2O + \Delta H = 672 \text{ kcal/mole}$$

because, in the process of cellular respiration, we require one molecule of water in order to utilize each atom of carbon in the glucose. Now each pair of H-atoms that becomes subject to the unknown processes of the mitochondrial cytochrome system provides the energy for forming three ATP molecules. Therefore, the mitochondrian is a power plant that essentially "burns" hydrogen to make water. We start with 12 pairs of hydrogen, and we get three ATPs for each pair, so each molecule of glucose makes 36 molecules of ATP.

Now we can make the connection to photosynthesis, because the above equation is obviously related to van Niel's equation for photosynthesis:

$$6CO_2 + 12H_2(X) \xrightarrow{h\nu} C_6H_{12}O_6 + 6H_2O + 12(X)$$

where (X) can be oxygen, sulphur, or some organic molecule. Green plants, in which

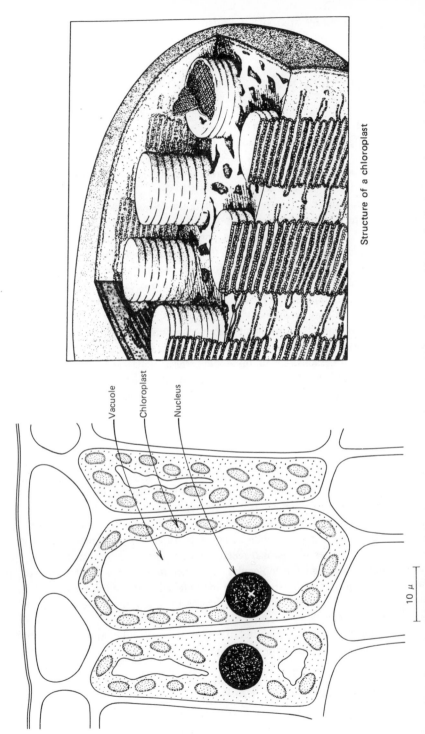

Structure of a chloroplast

Vacuole

Chloroplast

Nucleus

10 μ

Figure 9.3 The surface of a leaf is a single layer of epidermal cells. Just under this layer is a second layer of cells that are elongated and oriented with the long axis perpendicular to the epidermal layer; these elongated cells are called palisade cells. Air reaches the cells beneath the epidermal layer to the interior cells, or the mesophyll. The palisade cells are thus also mesophyll cells.

Palisade cells resemble animal cells in the presence of a cell membrane, a nucleus, mitochondria, and so forth. However, an extensive endoplasmic recticulum is absent, and it possesses three additional structures—the cell wall, vacuole, and chloroplasts.

the process is carried out by chloroplasts, represent the system with oxygen as the (X), and therefore we can write

$$6CO_2 + 12H_2O \xrightarrow{h\nu} C_6H_{12}O_6 + 6H_2O + 6O_2$$

This is the usual process one thinks of when photosynthesis is mentioned: the consumption of CO_2 and the production of O_2. Because the number of moles of O_2 produced equals the number of moles of CO_2 consumed, it is natural to guess that CO_2 is the "fuel" that supplies the O_2. This turns out not to be the case; the O_2 comes from "splitting" the H_2O. This is most easily shown by a tracer experiment. If we label the water with O^{18}, the tracer appears in the released oxygen; if we label the CO_2 with O^{18}, none of the released oxygen contains tracer.

Of course, the detailed mechanism of photosynthesis remains obscure, and the physical and chemical identities of many of the molecule complexes that play central roles are unknown, with the only evidence for their existence being the changes that occur in the absorption spectra at various stages of the process. We can summarize at least the crucial initial steps in the process, the so-called primary event by which the light energy needed for photosynthesis is captured.

In the higher plants, the photosynthetic mechanism is found in the chloroplasts, structures that are composed of a complex of membranes known as the lamellae. These membranes form sacks, called thylakoids; the lamellae are grouped to form a unit called a grana. Each grana is separated from the others by the stroma. The structure of a chloroplast is shown in Figure 9.3. The process of photosynthesis is, just like the formation of ATP in mitochondria, a membrane associated process. The thylakoids are the critical membranes, and they contain the pigment molecules that are the actual absorbers of the incident light. As you already know, the absorption spectra of organic molecules is usually much stronger in the ultraviolet than in other regions. Pigment molecules such as chlorophyll absorb strongly in the visible because they contain a particular ring compound, characteristic of the prophyrins. Attached to this ring is a hydrocarbon chain whose other end is anchored in the membrane. Of course, one should realize that there are a number of different forms of chlorophyll; on the basis of spectroscopic evidence some seven types are recognized at present. Light absorption in higher plants also receives a contribution from carotenoids. The molecular structure of pigment molecules is shown in Figure 9.4. The absorption bands of the chlorophylls are in the red and the blue, and that of the carotenoids is in the blue. Other photosynthetic systems have different pigment molecules and absorb in different wavelength regions.

What is the minimum energy that the pigment molecules must provide? The Gibbs free energy needed to drive the reaction is about

$$\Delta G \doteq 116 \text{ kcal/mole} \doteq 1.2 \ eV$$

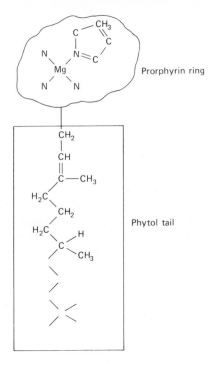

Prorphyrin ring

Phytol tail

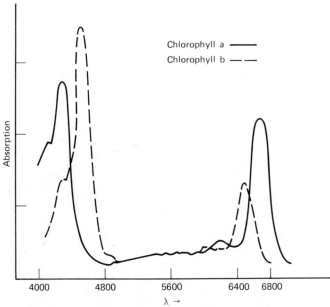

Chlorophyll a ———
Chlorophyll b — — —

λ →

Figure 9.4 The general structure of a chlorophyll molecule and the absorption spectra of two forms are shown.

Now the energy of a "mole" of photons at the peak of the red absorption band of chlorophyll is

$$\Delta G = 41 \text{ kcal}$$

and in the blue peak

$$\Delta G \cong 65 \text{ kcal}$$

Therefore, about two photons must be absorbed for each mole of carbon dioxide that is processed. In other words, there are probably two photochemical events. This is the heart of the problem; in some way the energy of the absorbed photons can be made available for work. The system that accomplishes this is the photosynthetic unit, a molecular complex including about 300 chlorophyll molecules. When a photon is absorbed by one of these, the molecule goes from ground to an excited state. The transition time for this is $\sim 10^{-15}$ sec. If the photon is from the red, the molecule is in the first excited state (singlet); if from the blue, the third excited state (singlet). The transition may begin and end on any particular rotational and vibrational level. The

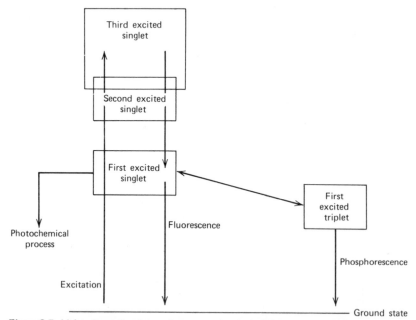

Figure 9.5 Light absorption by a pigment molecule is followed by a transition to one of several possible states. The only useful transition, as far as photosynthesis is concerned, is the one in which energy from the first excited state is transferred and becomes available to do photochemical work.

possible modes by which the molecules can then lose energy and return to ground are shown in Figure 9.5; obviously, the only important one for the plant is the mode that leads to energy transfer and the performance of chemical work.

A THERMODYNAMIC VIEW

In all of the cellular processes, animal and plant, the oxidation processes are usually thought of as those that break things down and, therefore, provide energy, in contrast to the reduction reactions that use energy. This is not really correct. If we want to oxidize an organic molecule, an electron or a hydrogen must be removed. This takes energy. On the other hand, a hydrogen or an electron can be taken by an organic molecule, producing reduction, and energy will be liberated. Thus, the work of the oxidation processes is paid for by the reduction processes and the cycle of ATP / ADP must be seen in that light.

In all of the processes an essential notion is the coupling between various reactions. We can described this, as Caplan has shown, by an appeal to irreversible thermodynamics. Suppose we have two processes and we identify one as the output or product and the other as the input. Then we can write the dissipation function as

$$\Phi = J_1 X_1 + J_2 X_2$$
$$= - \text{output power} + \text{input power}$$

The signs are chosen because the input must always be in the direction of its conjugate force, while the output will be against it. This is equivalent to saying that free energy is used up in producing the output. Now

$$\Phi \geqslant 0$$

so

$$J_1 X_1 < J_2 X_2$$

The efficiency may then be defined as

$$G = - J_1 X_1 / J_2 X_2$$

and physically is the fraction of the entropy production by the input process that is used to drive the output process. Caplan then introduces the degree of coupling, expressed in terms of coefficients connecting the fluxes and forces:

$$g = \frac{L_{12}}{\sqrt{L_{11}L_{22}}}$$

Let

$$z = \sqrt{L_{11}/L_{22}}$$

Then

$$\frac{J_1/J_2}{z} = \left[\frac{L_{12}}{\sqrt{L_{11}L_{22}}} + z(X_1/X_2) \right] \Big/ \left[1 + \frac{L_{12}}{\sqrt{L_{11}L_{22}}} z\left(\frac{X_1}{X_2}\right) \right]$$

$$= g + z(X_1/X_2)/1 + gz(X_1/X_2)$$

The sign of g may be either (+) or (−); it is (+) if one flux "carries" another in the same direction and (−) if one flux is "exchanged" for another. The value of g is thus +1 or −1 for fully coupled processes, and

$$J_1/J_2 = \pm z \qquad \text{if} \qquad g = \pm 1$$

Now consider two states; in both the output is zero. Clearly, if $|g| < |1|$, input energy is required to maintain the stationary state. When $J_1 = 0$ and X_1 is constant, we have a static situation; when $X_1 = 0$, a level flow. An example of the first situation is the concentration gradient of ions, maintained by active transport; of the second, the contraction of muscle under zero load. Again, input energy must be used to maintain $X_1 = 0$. Now

$$\epsilon = - \frac{\left(g + z \dfrac{X_1}{X_2} \right)}{\left(g + \dfrac{1}{z} \dfrac{X_2}{X_1} \right)}$$

In both the above cases, the efficiency is zero because the output is zero. Therefore, there must be some ratio of X_1/X_2 that maximizes ϵ. Obviously,

$$\epsilon_{max} \simeq \frac{g^2}{(1 - \sqrt{1 - g^2})^2}$$

Clearly, near the static and level flow conditions, actual energy transduction becomes small, since $\epsilon \rightarrow 0$. What is the activity under these conditions?

1. Static case: *maintains* $\Delta\mu$ or other forces.

2. Level flow case: *transports* at maximum rate.

We therefore define two new efficiencies:

1. Static case efficiency $\epsilon_s = -X_1/J_2 X_2$

$$\equiv \frac{\text{force produced}}{\text{rate of expenditure of metabolic energy}}$$

2. Flow case efficiency $\epsilon_F = J_1/J_2 X_2$

$$\equiv \frac{\text{rate of transport}}{\text{rate of expenditure of metabolic energy}}$$

Now if X_2 is fixed:

1. ϵ_s rises as X_1 increases and is maximum when static level is attained.

2. ϵ_F rises as X_1 decreases ($|q| < |1|$) and is maximum when level flow is attained.

Therefore, the conditions of static head and level flow are the conditions at which the metabolic energy can be most efficiently utilized.

We can now write a general relation for the mitochondrial process. We identify the "fluxes" as

J_P: phosphorylation

J_H: H^+-flow

J_O: the oxidation of the substrate

Then

$$\Phi = J_P A_P + J_H \Delta\mu_H + J_O A_O$$

where the A's are the affinities and $\Delta\mu$ is the chemical potential difference. Consider only the steady state and then

$$J_P = L_P A_P + L_{PH}\Delta\mu_H + L_{PO}A_O$$
$$J_H = L_{PH}A_P + L_H\Delta\mu_H + L_{OH}A_O$$
$$J_O = L_{PO}A_P + L_{OH}\Delta\mu_H + L_O A_O$$

Now we can consider the high-energy intermediate argument, the so-called chemical hypothesis. In that case,

$$J_O \leftrightarrow J_P \text{ and } J_H \qquad g > 0$$
$$J_P \leftrightarrow J_H \qquad g < 0$$

For the chemiosmotic hypothesis,

$$J_P \leftrightarrow J_O \qquad \text{only by} \qquad J_H$$

and

$$J_H = J_H{}^1 + J_H{}^2$$

where

$$J_H{}^1 = L_H{}^1 \Delta\mu_H + L_{OH}{}^1 A_O$$
$$J_O = L_{OH}{}^1 \Delta\mu_H + L_O{}^1 A_O$$
$$J_P = L_P{}^2 A_P + L_{PH}{}^2 \Delta\mu_H$$
$$J_H{}^2 = L_{PH}{}^2 A_P + L_H{}^2 \Delta\mu_H$$

Also

$$J_H = L_{PH}{}^2 A_P + L_H{}^1 \Delta\mu_H + L_H{}^2 \Delta\mu_H + L_{OH}{}^1 A_O$$

Therefore, three different g's must be considered: g_{PH}, g_{OH}, and g_{PO}. Now $L_{PO} = 0$, that is, J_P and J_O would be totally independent if $\Delta\mu_H = 0$. Also if $A_P = 0$, $J_P = 0$; and if $A_O = 0$, $J_O = 0$.

Let us now create a level flow by using an uncoupler of the processes, such as valinomycin. Then

$$g = \frac{g_{PO} + g_{PH}g_{OH}}{\sqrt{(1 - g_{PH}^2)(1 - g_{OH}^2)}}$$

Since $g_{PO} = -g_{PH}g_{OH}$, $g = 0$, as we expected. At static head,

$$A_P = R_P J_P + P_{PO} J_O$$
$$A_O = R_{PO} J_P + R_O J_O$$

and

$$g = \frac{-R_{PO}}{\sqrt{R_P R_O}} = 0.96$$

$$(A_P)_{J_P=0} = -18.4 \text{ kcal/mole}$$

From this we predict the efficiency to be 56%; the experimental value is 44%.

A CONTRIBUTION FROM ARTIFICIAL MEMBRANES

One of the difficulties in analyzing these cellular energy tranduction processes is the small scale of the transducer. An important contribution to the problem of ATP production is due to Skulachev. He has focused on the chemiosmotic hypothesis of Mitchell. Skulachev has emphasized that this hypothesis does not really have to be connected to any particular biological system. Instead, the hypothesis merely states that the energy needed for ATP synthesis comes from the energy stored in an electric field. Consider a capacitor. That energy has been accumulated by moving charge against a potential gradient and we may write

$$dw = V \, dQ$$

But, since $Q = CV$,

$$dw = Q/C \, dQ$$

and

$$w = \tfrac{1}{2} CV^2$$

In the biological case, the capacitor is a definite structure, the membrane.

Skulachev therefore made artificial membranes. These membranes contained a particular protein, bacteriorhodopsin, which is a light absorbing molecule, obtained in this case from *Halobacterium halobium*. If the membrane-pigment molecule system is illuminated, Skulachev found that a potential difference would appear across the membrane. In a complete system, not yet produced, this potential, which represents available energy obtained by absorption of the photons, could be used to drive the synthesis of ATP.

THE METABOLIC REQUIREMENT

We close this discussion of energy transduction by pointing out that we can express the transduction requirement for an animal as a whole by measuring the energy required to exactly maintain the animal. This energy, per unit time, is the metabolic rate. How does this rate depend on the mass of the animal? Suppose we assume that most of the animal is composed of muscle tissue and that muscular movement is the dominant energy consumer. We now use a dimensional argument.

The mass of the animal can be seen as the sum of the masses of the cylindrical pieces of the animal, the limbs, and the trunk. Assume these are roughly comparable in scale. Then

$$\mathcal{M} = \sum_i \rho_i(\ell_i)(d_i/2)^2$$

$$\simeq \text{constant } \rho \, \ell(d/2)^2 \pi$$

where ρ_i, l_i, and d_i are the density, length, and diameter of each cylinder. For animals, $\rho \cong 1$; also, since $\pi/4 \approx 1$, $\mathcal{M} \cong \text{constant } \ell d^2$. Now the bearing strength of a cylinder that supports itself without flexing requires

$$\ell \sim d^{2/3}$$

and this is roughly true for the limbs of an animal. Thus

$$\mathcal{M} \approx \text{constant} \times d^{2/3} \times d^2 \approx \text{constant } d^{8/3}$$

The cross-sectional area is

$$A = \pi(d/2)^2 = \pi/4d^2 \doteq d^2$$

So

$$\mathcal{M} \approx \text{constant } A^{4/3}$$

or

$$A \approx \text{constant } \mathcal{M}^{3/4}$$

According to Hill, the power of a muscle is proportional to its cross-sectional area. That power must be supported by energy expenditure, and that energy must originate in the metabolism of the animal's cells. Thus

$$\frac{dE}{dt} \approx \text{constant } A \approx \text{constant } \mathcal{M}^{3/4}$$

or

$$\frac{dE/dt}{\mathcal{M}^{3/4}} \doteq \text{constant}$$

for all animals. This is roughly correct; the value of the constant is ≈ 90. The result is known as Kleiber's law, and the dimensional analysis was first given by McMahon.

chap. 10

Radiation Biophysics

"After the discovery of X rays, there were many scientists who, marvelling at the penetrating power, never tired of looking at images of the skeleton of their own hands. However, their enjoyment was soon dampened by the observation of peculiar changes in the exposed skin."

Although the above is a succinct statement of the origin of biophysicists' interest in the effects of radiation, it is obvious that we need to seek a quantitative explanation for the origin of those effects.

THE BASIC EXPERIMENT

The key experimental result is the dose-response curve. This is obtained in the following way. We select a system whose response to radiation we wish to investigate. This system could be a culture tube of bacteria, an enzyme solution, or a tissue culture. We prepare a number of samples from our chosen system and expose each to a different amount of radiation. We then assay each sample in order to determine the fraction of the systems in the sample that were unaffected by the radiation dose and whose function is indistinguishable from that of an unirradiated sample. For a sample of an enzyme, the function could be its ability to catalyze a given reaction; for a sample of bacteria, the unaffected fraction could simply be the fraction of the bacteria that could still grow and divide. The dose-response curve is then obtained by plotting the unaffected fraction as a function of the amount of radiation received by the sample; examples are given in Figure 10.1.

The amount of radiation delivered to the sample may be expressed in a variety of ways. The fundamental unit is the roentgen, that amount of X-ray or gamma ray radiation required to produce 1 esu of charge in 1 cm^3 of air at standard temperature and pressure (STP), which is the same as $1/3 \times 10^9$ C per 0.001293 g of air. Another often used unit of radiation dose is the rad, which is the dose that produces 100 ergs of absorbed energy per gram of target. Unlike the roentgen, which is restricted to electromagnetic radiation, the rad may be used for all forms of radiation, be it particle or electromagnetic. Another unit is the roentgen equivalent man or rem, which is the dose required to produce the biological effect of 1 rad of X ray from a 200-kV tube. Then we have

$$\text{rem} = \text{relative biological effectiveness} \times \text{rad}$$

or

$$\text{rem} = \text{rbe} \times \text{rad}$$

The value of the relative biological effectiveness (rbe) can be quite variable. The rbe for fast neutrons is about 1 if the effect is simply survival; if, on the other hand, we are considering the induction of opacity in the eye, the rbe is about 20.

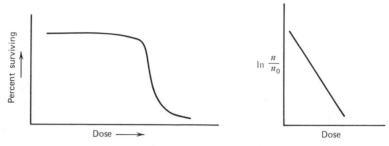

Figure 10.1 The dose-response curve.

TARGET THEORY

The dose-response curve is best analyzed in terms of the target theory. We begin by assuming that the events that occur upon irradiation, which we call the "hits," happen independently. Therefore, the Poisson distribution may be used to calculate the probability that a particular member of the irradiated sample will receive n hits. Thus

$$\text{Probability of } n \text{ hits} \equiv P(n) = \frac{(vD)^n e^{-vD}}{n!}$$

where v is the volume of the target, or precisely, the volume of the target that is sensitive to the effect of the radiation, and D is the radiation dose.

If a given member of the sample is inactivated by n hits, then all the members of the sample that receive fewer than n hits will survive and are assumed to be unaffected by the radiation. Therefore, the probability of survival is

$$P(0) + P(1) + P(2) + \ldots P(n-1)$$

The simplest definition of the probability is that it is the ratio of the number of events we are interested in to the total number of events we observed. Thus the probability of survival is

$$\frac{N_{\text{sur}}}{N_0} = P(0) + \ldots P(n-1)$$

$$= e^{-vD}\left(1 + vD + \frac{(vD)^2}{2!} + \ldots \frac{(vD)^{n-1}}{(n-1)!}\right)$$

$$= e^{-vD} \sum_{0}^{n-1} \frac{(vD)^k}{k!}$$

If $n = 1$, then

$$\frac{N_{sur}}{N_0} = e^{-vD}$$

It is also possible to derive expressions for systems with several targets; this is the case in which a given member of the sample must receive n hits on each of several different sensitive volumes within the given member in order to be inactivated. However, such situations are complicated, and the inherent variations in the dose-response curves limit the accurate interpretation of the equations.

An alternate approach attempts to deal with the inherent variation in the following way. Let

$$\frac{N}{N_0} = 1 - w(D)$$

where $w(D)$ is the probability that a dose D will produce an effect. The value of $w(D)$ will range from 0 to 1, and clearly each hit will increase the value of $w(D)$. Now let

$$w(D) = \frac{dw(D)}{dD}$$

We recall from statistics the definitions of the variances:

$$m_1 = \int_0^\infty Dw(D)dD$$

$$m_2 = \int_0^\infty D^2 w(D)dD$$

and

$$\sigma = m_2 - m_1{}^2$$

The steepness of the dose-response curve may then be defined as

$$S = m_1{}^2/\sigma^2$$

From this, it follows that $\bar{n}$, the average number of hits needed to produce an effect, must be limited by the condition that

$$\bar{n} \geqslant S$$

Thus we can determine the value of S from experiment, which gives us the least number of hits required to describe the effect of the radiation. Fortunately, much of the experimental evidence suggests that the dominant processes are single hit events.

PROCESSES OF RADIATION ABSORPTION

We now consider the processes by which the energy carried by the radiation is actually absorbed. As you already know, the absorption of electromagnetic radiation by a material generally can be described by a Lambert-Beer law:

$$I(x) = I_0 e^{-\mu x}$$

We wish to know what the specific processes are that contribute to the total absorption coefficient μ. In the case of X rays and gamma rays, there are three specific mechanisms: the photoelectric effect, the Compton effect, and pair production. These mechanisms may be summarized as follows:

1. *The photoelectric effect.* The energy of the incident photon may be greater than the binding energy of the electron to the atom or molecule. If it is, the energy of the photon is given up in doing the work needed to remove the electron and to ionize the atom or molecule. The photoelectric effect contribution to the overall absorption rises rapidly with the atomic number, going about as Z^4.

2. The Compton effect is that process where an X-ray of γ-ray photon collides with an electron and in so doing gives up part of the energy of the photon, thereby producing an electron with increased kinetic energy and a photon of decreased frequency. The energy that can be transferred to the electron is not necessarily large; a 10-KeV photon always retains at least 95% of its initial energy. However, the possibility of large energy transfers does increase with the energy of the incident photon. Since the Compton effect is essentially a photon-electron collision, its contribution to the overall absorption coefficient goes up with the number of electrons with which collisions might occur and so with Z.

3. The pair production process is the formation of an electron-positron pair from a photon with an energy greater than the sum of the rest mass

energies of the two particles, which is about 1 MeV. The process is the direct conversion of energy to mass, but it must occur in the presence of a third body, usually a nucleus, in order to maintain conservation of momentum. The contribution of this process to overall absorption increases as Z^2.

We also need to consider the mechanism of absorption for particles. In the case of neutrons, the absence of charge means that they are unaffected by any Coulomb repulsion from the nucleus. Generally, the lower the energy of the neutron, the more likely it is to be captured. The capture of a neutron by a nucleus produces a nucleus in an excited state; the excited nucleus decays to ground by the emission of a γ-ray in the case of light nuclei and by the emission of a γ-ray and a proton or α-particle in the case of heavier nuclei. The biological effects of neutron irradiation are mainly due to this secondary radiation, as well as to the recoil of the target nucleus in the capture process, which occurs because of conservation of momentum.

In the case of charged particles, we can understand the effects by either a particle collision model or in terms of the field of the particle exerting a force on other charged particles, principally electrons, in the target. Using this latter view, Bloch and Bethe showed that

$$-\frac{dE}{dx} \cong \frac{4\pi(ze \times e)^2}{m_e v^2} N \left[\ln \frac{2m_e v^2}{I_0} - \ln(1 - \beta^2) - \beta^2 \right] Z$$

for the case of particles much heavier than electrons. In the above equation, note that the mass of the bombarding particle does not enter. Also the average ionization potential may be approximated by

$$I_0 \approx 13.5 \, Z \, (\text{eV})$$

The above processes do not produce uniformly ionized material. Instead, there is a distribution in the number of ionizations produced in an irradiated sample and this distribution is given by

$$dn = \frac{2e^4 z^2 N Z_1}{m v^2} \left(\frac{1}{w_1} - \frac{1}{w_2} \right)$$

where

z is the charge on the fast particle

Z_1 is the charge on the target

w_1 and w_2 are the upper and lower energy ranges

The above equation shows that the energy loss is proportional to the number density of atoms in the target multiplied by the number of electrons per atom, Z. Since that is roughly the same for a variety of biological materials, it follows that the energy loss is roughly the same for all these materials.

The energy loss per unit length, the dE/dx in the Bloch-Bethe equation, is often called the linear energy transfer, or LET; when divided by the density of the target material, it is called the mass stopping power.

How is the energy of the passing charged particle transferred to the target molecule? To answer this question, we need first to define the oscillator strength of f-value, such that

$$\sum_{\substack{\text{all} \\ \text{states}}} f_s = Z$$

Then the optical absorption coefficient μ is proportional to f at a particular frequency, s. If the transition is within the continuum,

$$\mu(v) \sim df/dv$$

Thus a molecule exposed to a continuous spectrum in which there are equal numbers of photons in each frequency interval, say from the X-ray region to the visible, will be excited to a state s with a probability f_s. In other words, the number of molecules excited to that state will be given by

$$N_s = \text{constant} \times f_s$$

Now consider a passing charge particle. The force acting on the molecular electrons and in the direction of the particle's motion can be written

$$F_{TRANS} = \sum k_i \cos 2\pi v_i t$$

Each frequency has a constant energy:

$$I(v) = \text{constant}$$

The number of photons in that interval is

$$n(\nu) = \text{constant}/h\nu$$

Thus the passage of a charged particle is equivalent to an exposure to a continuum with a frequency distribution of

$$\sim 1/h\nu$$

So since

$$1/h\nu_s = 1/E_s$$

therefore

$$N_s \sim f_s/E_s$$

or, for a continuous spectrum,

$$N(E) \sim \frac{df}{dE}/E$$

Since f is proportional to the number of electrons in the shell where the activation occurs and, since, for the atomic case, f increases with the distance from the nucleus, it follows that events are primarily concerned with the valence electrons. In molecules, most of the range of f-values are for frequencies whose corresponding energies are above the ionization potential of the molecule. Since the excitation is independent of either the type of particle or its velocity, all ionizing radiation produces, qualitatively, similar effects when discussed from the molecular level. The one exception to this rule is for radiation in the vacuum ultraviolet; in that case specific high-energy excitations may occur. Hence, we say that the passage of a charged particle can generate electrons from atoms or molecules by either the so-called "knock-on" collision, in which a fast particle collides with an electron that is relatively speaking, stationary, or by a glancing collision, in which the particle perturbs the atom or molecule with its electric field at distances of up to about 1 nm. The glancing collisions are roughly an order of magnitude more frequent than the knock-on collisions.

ACTION OF RADIATION

The preceding material permits us to begin to understand why radiation can be such a potent agent. The reason for this is initially somewhat obscure, because the energy delivered by such means seems quite small. We recall that a dose of one rad delivers 100 ergs to a gram of material. The maximum increase in temperature that could be produced by this dose is less than 2×10^{-6} °K. Yet the effects of a 100 rad dose on a variety of samples is detectable, and a dose of about five times that amount, delivered over the whole body, would very likely kill a human. We can begin to understand the potency of radiation by realizing that the energy of a particle is essentially delivered on a molecular scale. For example, suppose we consider a single macromolecule of intermediate weight. As a result of irradiation, this molecule absorbs 100 eV. Then, since the mass of such a molecule is approximately

$$m \approx 2 \times 10^{-20} \text{ g}$$

and the energy absorbed is approximately

$$E_{abs} \approx 4 \times 10^{-18} \text{ cal}$$

we can calculate the effective temperature increase of the molecule by assuming a specific heat of approximately 1 cal/g °C^{-1}. We obtain

$$\Delta T = \frac{E_{abs}}{m} \approx 200 \text{ °C}$$

Another way of looking at the process is to realize that each collision event associated with the passage of a fast particle delivers about 60 eVs; that is sufficeint to generate free electrons, which will themselves have enough energy to produce secondary electrons by collisions with the surrounding molecules. Since the average binding energy between the atoms in the molecule is only about 3 to 4 eV, the mean energy per primary event is some 20 times the energy of the chemical bond. Pollard has given an apt image for the process; the action of ionizing radiation is like a bolt of lightning and wherever it strikes, regardless of the target, the action will be drastic.

The idea of a hit can now be sharpened in the following way. First we note that the target approach leads to a measure of the volume of the radiation sensitive structure in the system. We recall that for the case of a response due to a single hit ($n = 1$), we have

$$N/N_0 = e^{-vD}$$

Therefore, if we plot ln N/N_0 versus D, the target volume is simply the slope of the line. Since $vD = 1$ when $N/N_0 = 37\%$, the target volume is also the reciprocal of D_{37}, the dose at which 37% of the targets survive, or are unaffected. Another way of saying this is that at D_{37}, the total number of hits is the same as the number of objects irradiated. Therefore, if we measure D_{37} for a sample, say in ergs/g, and divide by the mean energy per hit, we obtain the number of hits per gram; that is, the number of hits per gram is the same as the number of objects per gram.

We can express this in a quantitative way:

$$1 \text{ rad} = 100 \text{ ergs/g} = 6.24 \times 10^{13} \text{ eV/g}$$

If the mean energy/primary ionization is 60 eV, then

$$1 \text{ rad} = \frac{6.24 \times 10^{13}}{60} = 1.04 \times 10^{12} \text{ hits/g}$$

The mass of the target in grams is thus

$$m = 1/D_{37} \times 1/1.04 \times 10^{12} = 0.96 \times 10^{-12} \frac{1}{D_{37}}$$

Since m/v = the density, ρ, we have

$$v = m/\rho = 0.96 \times 10^{-12}/\rho D_{37} \text{ cm}^3$$

This equation can be tested by determining the radiation dose needed to inactivate enzymes of known molecular weight. If we multiply the right-hand side by Avogodro's number, we obtain the molecular weight of the target. Molecular weights of a number of enzymes have been calculated from measurements of D_{37} for those enzymes. In general, there is a very good agreement for enough cases to make one confident of the validity of the equation. Unfortunately, a variety of complications and uncertainties make it unlikely that molecular weights determined by target theory can be used to assay macromolecules that cannot be obtained in the forms required for molecular weight determinations by standard methods.

The notion of the target must not be overinterpreted. Damage to the "target" by "hits" from the radiation is certainly a reality, but action also occurs due to radiation effects on the surroundings. This situation has led to the notion of direct and indirect action.

Although it was once believed that indirect effects could occur only with "wet" material, this is now known to be incorrect. However, the indirect process in solutions can be rather well described, and therefore we will limit the discussion to that case.

The essential solvent is water. The events occurring upon radiation absorption can be described by the following chemical reactions:

$$H_2O \xrightarrow{12.56 \text{ eV}} H_2O^+ + e^-$$

so

$$H_2O \longrightarrow H^+ + OH$$

and

$$e^- + H_2O \longrightarrow OH^- + H$$

Also

$$H_2O \longrightarrow H_2O^* \longrightarrow H + OH$$

where H and OH are called water radicals. The electron may become stabilized as a hydrated electron, e_{aq}^-, and in this form it can produce effects at a considerable distance from its point of origin because it can diffuse through the water for considerable distances. We define the number of products formed per 100 eV absorbed as the G-value. (The range of pH is restricted to $3 < pH < 10$.) The G-values for the above products are

$$G_{e^-, \text{aq}} = 2.3$$
$$G_H = 0.6$$
$$G_{OH} = 2.3$$

Recombination processes lead to the formation of H_2, H_2O_2, and H_2O. The first two are clearly potential sources of damage to biological structures. High LETs will favor the formation of these products because the local concentration of radicals is high. This is the explanation for the fact that X rays are more effective than α-particles in the production of enzyme inactivation.

The indirect action in solution can be analyzed from either the assumption that only a single reaction with a target molecule is possible, or from the assumption that the reaction is random and thus the probability of an interaction with a previously "hit" molecule is the same as the probability of an interaction with an unhit target. In the first case,

$$\frac{N}{N_0} = 1 - kD$$

In the second, which is a likely case for large biomolecules,

$$\frac{N}{N_0} = e^{-kD}$$

The first possibility is in fact observed for small molecules and is the dominant mode in such reactions as

$$FeSO_4 \rightarrow OH \rightarrow Fe^{2+} + OH \rightarrow Fe^{3+} + OH^-$$

a reaction often used in dosimetry calibrations.

In dilute aqueous solutions, indirect action is the dominant mode of damage. This can be seen from the results of radiation inactivation studies of ribonuclease (RNase). In solution, with a concentration of 5 mg/ml, RNase has $D_{37} = 0.4$ Mrad, while when dry, $D_{37} = 42$ Mrad. Since the target volume is inversely proportional to $1/D_{37}$, we have

$$\frac{v_{\text{in solution}}}{v_{\text{dry}}} = D_{37}^{\text{dry}}/D_{37}^{\text{solution}} = 42/0.4 = 105$$

which means the sensitive volume in solution is 105 times larger than the sensitive volume in the dry state. This suggests that in dilute solutions, nearly all of the RNase molecules are inactivated by radiation products formed in the water. In order to be sure of this, we must eliminate the possibility that the energy required to inactivate the molecule does not decrease when it is in solution. Recall that

$$G = \text{number inactivated}/100 \text{ eV absorbed}$$

$$= \frac{Z_m}{Z_t} \quad \frac{\text{the number of irradiated molecules/gram}}{\text{the number of 100 eV units/gram at } D_{37}}$$

In the case of a dry target,

$$Z_m^{\text{dry}} = N_A/\text{mol wt} = 6.022 \times 10^{23}/\text{mol wt}$$

Since 1 rad = 100 ergs/g = 6.24 x 10^{13} eV/g, we have

$$Z_t = 6.24 \times 10^{11} D_{37}$$

Thus for the dry case,

$$G_{dry} = \frac{9.65 \times 10^{11}}{D_{37}(\text{mol wt})}$$

In solution,

$$Z_m^{wet} = \frac{6.002 \times 10^{23}}{\text{mol wt}} \times \text{concn}$$

$$G_{wet} = G_{dry} \times \text{concn } l$$

For RNAse, with molecular weight of 13, 680, we find

$$G_{dry} = 1.7$$

and

$$G_{wet} = 0.9$$

if the concentration is 0.005. Given that the ratio of the D_{37} values for these cases is ~ 100:1, the two G values are close enough so that even if the activation is somewhat changed in solution, the change is only a minor contribution to the inactivation process.

All of the above matters play a role in understanding the effect of radiation on a population of cells. Consider an example first given by Pollard. Suppose we irradiate a colony of *E. coli* bacterial cells. A typical *E. coli* cell is a cylinder about 3 μ long with a radius of 1 μ and a mass of 10^{-12} to 10^{-13} g. It is 70% water and contains:

1. 20,000 molecules of DNA with a molecular weight of ~ 10^7.

2. 40,000 molecules of RNA with a molecular weight of ~ 10^6.

3. 5×10^6 molecules of protein with a molecular weight of ~ 10^5.

4. 3×10^7 molecules of lipid with a molecular weight of ~ 10^{3-4}.

5. 1.6×10^7 molecules of phospholipid with a molecular weight of ~ 10^{3-4}.

The total cell volume is about 2.25×10^{-12} cm^3, so the volume fractions of the major components are:

1. Nucleic acids: $\sim 10^{-6}$ of the cell volume
2. Proteins: $\sim 10^{-7}$ of the cell volume.
3. Lipids: $\sim 10^{-9}$ of the cell volume.

Thus, the chance of inactivating a particular molecule is obviously not large. For a dose of 1 kr, the probabilities are:

1. Nucleic acid: $\sim 7 \times 10^{-4}$.
2. Protein: $\sim 8 \times 10^{-5}$.
3. Lipid: $\sim 10^{-6}$.

This means we inactivate about a dozen molecules of DNA; two dozen of RNA; about 400 protein; 30 lipids; and 20 phospholipids. To these, which are inactivated by direct action, we must add those subjected to indirect action. The degree of indirect action is presumably dependent on the area of the molecule that can be attacked. The surface areas of the various molecules are:

1. DNA: $\sim 5 \times 10^{-7}$ cm^2.
2. RNA: $\sim 7 \times 10^{-7}$ cm^2.
3. Protein: $\sim 6 \times 10^{-6}$ cm^2.
4. Lipid: $\sim 2 \times 10^{-6}$ cm^2.

If 70% of the cell is water, indirect action is obviously going to be important. Assume 1000 R produces about 1000 primary events. If the probability of inactivation is directly proportional to the surface areas of the various molecules, the total number of molecules inactivated by both direct and indirect action is:

1. ~ 60 DNA molecules.
2. ~ 120 RNA molecules.
3. ~ 600 protein molecules.
4. ~ 225 lipid molecules.

From our knowledge of molecular biology, we suspect that the hits on the nucleic acid component are the major contributors to the inactivation of the cell. Our crude calculation suggests that about 100 cells per 1000 should be inactivated; the actual experimental result is ~ 15 to 20% per 1000 R.

The hypothesis that the target is the nucleic acid component is supported by the observation that the effects of radiation on cells and organs are all delayed effects. Also note that radiation effects on *E. coli* colonies are observed at doses of 1000 R, whereas D_{37} for enzyme inactivation can be several hundred times that. Other experimental results interpreted by target theory consistently suggest that the process of cell inactivation centers on the nucleic acid. For example, in *Drosophila*, a dose of about 4000 R is about D_{37}. Therefore, the target volume is about $v = 4.1 \times 10^{-16}$ cm^3, which is equivalent to a molecular weight of 3.2×10^8, approximately correct for DNA. This is supported by the incidence of mutations. In flies that survive 4×10^3 R, the incidence of mutation is 1:500. We know that mutations are produced by changes with a particular region of the DNA, which is the gene. If dn is the number of mutated flies per n cases, we would expect

$$dn/n = v \, dI$$

From the data,

$$v = 4.1 \times 10^{-19} \text{ cm}^3 \equiv \text{mol wt} = 320,000$$

Since other crossing experiments suggest that there are about 3000 genes, the total molecular weight of the genetic material must be about 9.6×10^8, which, given the simplicity of the argument, must be considered in good agreement with the above result.

These simple arguments let us see how to analyze the effect of the radiation on the colony of *E. coli* as a whole. To stop the colony from growing, we must hit all of the targets, each of which has a sensitive volume equal to the nucleic acid volume. The probability of a hit is

$$P(\text{hit}) = 1 - e^{-vD} \sum_{0}^{n-1} (vD)^k/k!$$

Since the cells must be hit, the probability of getting m is given by

$$[P(\text{hit})]^m = \left[1 - e^{-vD} \sum_{0}^{n-1} (vD)^k/k! \right]^m$$

Since the most likely case is $n = 1$,

$$\frac{N \text{ sur}}{N_0} = 1 - (1 - e^{-vD})^m$$

which is

$$\frac{N \text{ sur}}{N_0} \doteq 1 - (1 - me^{-vD} \pm \ldots e^{-mvD})$$

$$\doteq me^{-vD}$$

A conclusive proof that the DNA component is the sensitive target can be assembled from several lines of argument:

1. In general, the relative sensitivity of bacteria to radiation falls with the increasing adenine-thymine content of the bacterial DNA. Such a direct relation is practically proof in itself.

2. It is possible to sensitize nucleic acid directly by the use of certain base analogues. The most common analog is bromouracil (BU), which can be incorporated into DNA in place of thymine.

Sensitization by BU is best shown in phage DNA. This is accomplished by growing the host bacteria under conditions that prevent thymine synthesis. If these cells are transferred to a medium with BU, phages grown in these cells then take up the BU and incorporate it into their DNA. BU sensitizes this DNA to direct action by radiation, as is shown from the fact that irradiation of aqueous solutions of BU is without effect. In the case of bacteria, we observe that the sensitivity to radiation increases with the BU content of the DNA. The easiest experiments are done with mutant bacteria that are naturally deficient in thymine and readily incorporate BU into their DNA.

The damage to the DNA is in the form of actual breaks in the strands. This has been shown directly. In the spider wort, *Tradescantia*, α-particle irradiation was followed by direct microscopic examination of the cells. If we plot a dose-response curve using the percent of intact nuclei as our criterion, we find the predicted linear relation between $\log N/N_0$ and D. The area of a cell nucleus is $\sim 11.3 \times 10^{-7}$ cm^2; the cross section for breaking the nucleic acid strand is found to be about half that. The total strand length is 972 μm. The effective thickness of the strand is thus ~ 80 nm as deduced from the radiation result; the known thickness is ~ 100 nm.

A second way to show that actual breakage occurs is to use the BU substituted cells. After irradiation, the cell walls are borken up chemically and the DNA isolated by centrifugation. Irradiated DNA will be broken and will be sedimented less readily than will whole DNA or longer strands. It is observed that BU substituted DNA shows about three times the damage rate of unsubstituted DNA, and these differences are consistent with the differences in radiation sensitivity shown by the two cell populations.

In the case of bacteria, it is possible to define the nucleic acid damage process a bit more precisely. This is done by "pulse" irradiation experiments with carbon-14 labelled uracil. As you recall, uracil is a normal component of RNA, but not DNA. Thus a bacterium that is briefly exposed to uracil will almost immediately incorporate that uracil into its RNA, and particularly into mRNA. The brief exposure is best accomplished by simply giving the cell a bit of labelled uracil and then flooding it with unlabelled uracil to end the pulse. These experiments clearly show that in cells that are irradiated before being give a pulse, very little labelled uracil appears in the mRNA, which is in very marked contrast to the results for unirradiated cells. The obvious conclusion is that radiation damage is localized on the transcription stage. This result, and several others, permit us to break the nucleic acid damage process into parts: For a given dose, the mRNA system is the most fragile, the DNA synthesis process is the least, and the DNA itself falls somewhere in the middle.

Up to this point we have been discussing the effect of radiation as if the radiation dose delivered to the target were the only variable. In fact there are three other important factors that should be considered. They are temperature, oxygen concentration, and repair action.

In many systems, the sensitivity to radiation is observed to decrease with temperature reduction. Presumably, the temperature effect results from the dependence of a rate constant on temperature, and the effect is apparently due to the action of light radicals that are formed in either wet or dry material. If that were so, then the rate constant, k, should be given by a constant $\times e^{-E/RT}$ where E is the activation energy of the reaction. In fact, such an equation does fit a number of different experimental results. However, there are systems that are not described by the results from such a simple assumption and unsolved problems remain.

The oxygen effect was discovered by Thoady and Read, who observed in the course of experiments with *Vicia faba* that the exclusion of oxygen reduced the frequency of chromosome breaks in the cells of that legume by a factor of 2 to 3. The origin of the oxygen effect remaisn a difficult and complex problem, but some progress has been made, based mainly on a suggestion, due originally to Flanders, that damage could be divided into two categories. The first type was known as light damage and was described as that damage due to low-energy deposition. The heavy damage was that produced by the rarer large amounts of energy. By considering the shape of the curve produced by plotting $1/D_{37}$ versus the linear energy transfer, it is possible to show that the oxygen sensitization declines with increasing LET because the frequency of heavy damage is rising. Therefore, the oxygen effect presumably has its origin in the light damage category. This suggests a variety of chemical problems and matters associated with the repair of radiation damage.

The repair of radiation damage is a complex problem. A repair system does exist that mends, or attempts to mend, breaks produced in the nucleic acid strands by radiation. This system was discovered by Elkind and Sutton, who found that in a number of cell cultures the effect of a single dose on a colony was greater than the effect of two doses given separately, but with the total dosage as before. Understanding this process was initially complicated by the fact that for colonies of synchronous cells, the radiation sensitivity of DNA varies over the cell cycle. However, considerable progress has been made and it now appears, based on the work of Johansen and Boyle, that the oxygen effect does not particularly involve the repair system.

chap. 11

Salt And Water Transport In The Gut

In all animals except the *Pogonophora*, which are deep sea marine worms, the absorption of nutrients takes place in the gut; in the *Pogonophora*, which lack a gut, absorption of nutrients takes place directly through surface tissue. In higher animals, the main products formed by breaking down the carbohydrates, proteins, and fats enter the circulation of blood and lymph by crossing the epithelium of the small intestine, which is a single layer of cells, though much convoluted. The epithelium is thus the dividing layer between the inside of the gut (the mucosal side or lumen) and the blood and lymph (serosal side).

STRUCTURE OF THE GUT

The gut is a complicated structure whose central feature is the morphological development required to maximize the surface area. Four developments have occurred to maximize the surface area. The first is the infolding of the tissue, which increases the surface area by about a factor of three over that of a simple cylinder of the same exterior dimensions. From these infolded sheets, smaller structures arise. Called villi, these projections are about $\frac{1}{2}$ to 1 mm long and occur at the rate of ~ 25 villi/ mm^2 of tissue; the total area of the gut is now about 30 times that of a simple cylinder. Finally, from each villi project numerous smaller structures, the subcellular microvilli, whose scale is ~ 1 μm. The effective area is now increased by a factor of ~ 600. Thus, although a piece of gut about 6 m long would have a surface area of ~ 1 m^2 if it were just a tube, in actuality the area is about 600 m^2. The structures are shown in Figure 11.1.

The epithelial cells are highly differentiated, as befits their role in the complex processes of absorption and transport. A typical absorptive cell is shown in Figure 11.2.

The microvilli are found under electron microscopy to show the characteristic "rail-road track" of the bilayer membrane. If the electron microscopy results are reliable, the microvilli membranes are somewhat thicker ($\sim 9.5-11.5$ nm) than is the membrane in the portion of the absorptive cell that points away from the gut center.

Let us summarize some of the processes occurring in the gut. Carbohydrate intake is 200 to 800 g per day. The enzymes in the saliva and in the pancreatic juice that are released into the lumen catalyze the hydrolysis of starch and glycogen. This is a very rapid process; starch is reduced to small molecules in about 10 minutes. Roughly 15% will be reduced to hexoses, and the remainder will be dissacharides. The dissarcarides are then attacked by different enzymes:

Sugar + sucrase $\longrightarrow$ glucose + fructose + sucrase

Lactose + lactase $\longrightarrow$ glucose + glactose + lactase

$\left. \begin{array}{l} \text{Starch} \\ \text{Glycogen} \end{array} \right\} \longrightarrow$ dissacharide + maltase $\longrightarrow$ glucose + maltase

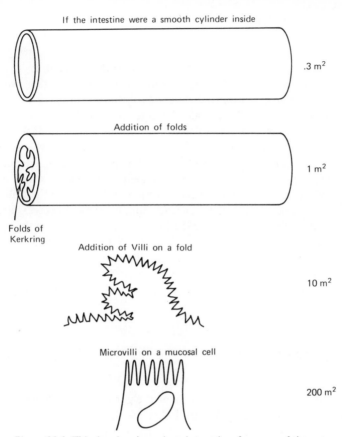

If the intestine were a smooth cylinder inside

.3 m²

Addition of folds

1 m²

Folds of
Kerkring

Addition of Villi on a fold

10 m²

Microvilli on a mucosal cell

200 m²

Figure 11.1 This drawing shows how internal surface area of the gut is increased by complex folding and the formation of projections.

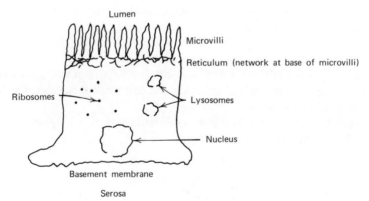

Lumen

Microvilli

Reticulum (network at base of microvilli)

Ribosomes

Lysosomes

Nucleus

Basement membrane

Serosa

Figure 11.2 A typical absorption cell from the gut.

The enzymes for these processes are localized in the microvilli. Dissacharides in the lumen diffuse into the microvilli, are broken down to hexoses, and enter the blood and lymph circulation. Thus the epithelium is a transporter of hexoses obtained by digestion of hexose polymers, be they very large such as starch or glycogen, with several thousand monomers, or very small indeed, such as the two monomer "polymers," sucrose and lactose. Carbohydrate digestion is the process that produces the hexoses (e.g., glucose, galactose and fructose) that can be transported across the gut. Of these hexoses, glucose and galactose are actively transported; all other diffuse. In the case of the hexose pump, ATP is of course the energy source and, not surprisingly, Na^+ is involved; each $ATP \rightarrow ADP$ step moves several neutral glucose molecules and Na-ions. The pump is on the lumenal side, and its effect is to greatly increase the concentration of the actively transported hexoses in the cells. Since this concentration is greater than the concentration of these hexoses in the blood, diffusion into the serosal side of the epithelium will occur.

In the case of proteins, the process is similar; the pancreas releases enzymes into the lumen to produce protein hydrolysis and the reduction of protein to peptides. Enzymes in the microvilli reduce these to single amino acids, and these move to the serosal side, in at least three cases by active transport, as well as by diffusion.

ANALYZING TRANSPORT

The passage of materials across the walls of the gut is clearly a membrane transport process. Thus, it is reasonable to expect that we should be able to use theoretical methods for membrane processes to analyze that transport. One of the most essential aspects of transport in the gut of higher animals is the salt and water exchange. In the human intestine, this exchange is quite impressive: ~ 8 liters of water and 1 mole of sodium chlorine per 24 hours. Where does the water come from? Although some is taken in as liquid, either directly or in food, most of the water involved in the transport arises through secretions that occur along the digestive tract. These secretions are necessary both for lubrication of the tract and for the transport of the various enzymes required in the digestive process.

Let us analyze the salt and water transport. It is important to begin by realizing that the flows are not the result of diffusion and osmosis, although such statements can be found in some textbooks.

We recall that the force that acts on a substance moving in only one direction in a system without temperature or pressure gradients is

$$\frac{-d\bar{\mu}}{dx} = -RT \frac{d \ln a}{dx} - zF \frac{d\psi}{dx}$$

and the flux is

$$J = -uC\frac{d\bar{\mu}}{dx} \text{ moles/sec}$$

where

u = mobility in cm/(sec) (unit force)

μ is the electrochemical potential

a is the chemical activity

Since the net flux of a substance is the difference between the flux from side one to side two and that from side two to side one, we can write

$$J_{net} = J_{12}\left(1 - \frac{a_{12}}{a_{21}}e^{ZF\,\Delta\psi/RT}\right) \tag{11.1}$$

where $\Delta\psi$ is the potential difference. This equation shows that:

1. The flux of an uncharged substance acting only under diffusion is determined by the chemical activities on each side of the membrane.

2. The flux of charged substance is given by the above equation; the potential retards the flow in the direction in which the potential has the same sign as the particle.

If a substance does not obey the above equation, it simply means that the electrochemical potential difference is not the only force acting or that some other phenomena must be considered. Let us extend the discussion of membrane processes in Chapter 6 to consider some of these other phenomena.

The most obvious example of the existence of an additional force arises from the case when a *net* volume flow occurs across the membrane. Although the process may be described by diffusion, it is a direct example of coupling; the flow of solvent influences the flow of solute and leads to the phenomena known as solvent-drag. This effect is important in porous membranes. It can be detected in frog skin and bladder membranes by placing equal concentrations of thiourea on either side of a membrane preparation. If there is no volume flow of water, the fluxes are equal; therefore, the transport of thiourea is passive. Nevertheless, if a net water flow is produced by an osmotic gradient, a net flow of thiourea occurs in the same direction. Since the fluxes do not obey Equation 11.1, some would say that active transport is present. Clearly, that is not the case, because no metabolic energy is actually required. It can be shown that for a single solute at equal concentration, C_i, on both sides, the solvent-drag produces a net solute flow, J_i, given by

$$J_i = C_i J_v - \sigma_i C_i J_v$$

where J_v is the volume flow of solvent and σ_i is a coefficient that can be evaluated.

Another way in which deviations from Equation 11.1 may occur is if the molecules or ions cannot freely move across the membrane because only certain pathways or channels are available for movement. In a sense, there is a traffic jam; an example of this is the so-called single-file diffusion proposed for the transport of potassium in nerve fibers under certain conditions and the movement of chloride in muscle tissue. Again, although Equation 11.1 would not predict correct fluxes, no active transport would be occurring.

Of course, real active transport, involving the actual expenditure of metabolic energy, does occur. Other processes such as exchange diffusion or the formation of various chemical complexes between the membrane proteins and the substances being transported also can confuse the issue. Hence in an investigation of NaCl and H_2O transport by the cells of the intestinal epithelium, the first requirement is to decide if either NaCl or H_2O or both obey Equation 11.1. Since NaCl is essentially totally dissociated into Na and Cl ions, they must be considered as separate species.

Tracer experiments show that Na-ions can cross the intestinal wall in either direction. Furthermore, the transport rate of Na is not equal to that of Cl. Since measurements of the Na flux also show that transport against a concentration gradient is possible, it becomes essential to obtain reliable values of the electrical potential difference across the membrane. In the small intestine, it is found that $\Delta\psi \simeq 5 - 12$ mV and in the large intestine, $\Delta\psi \simeq 5 - 50$ mV. In both, the positive potential is in the direction of serosa to mucosa. From these measurements, Curran was able to show that the net Na-flux did not obey Equation 11.1. For example, with a concentration of Na maintained equal to 140 mM on both sides of a membrane of intestinal tissue from a rat, and with the potential difference equal to $+ 7.2$ mV in the serosa, the net Na-flux is predicted to be from the serosa to the mucosa. In fact, a flux is observed in the opposite direction.

If the two components of the flux are now studied separately, it is found that changes in the net flux across the membrane are due solely to variations in the mucosal to serosal flux and that the serosal to mucosal flux is approximately constant. The value of this latter component is correctly predicted by Equation 11.1, but of course the former component is not.

These conclusions can be tested by calculating the potential required to produce an electrodiffusion from the serosa to the mucosa that will exceed the mucosal to serosal flux. This potential is found to be about 15 to 20 mV and, if it is applied to an intestinal membrane, the net Na-flow is indeed observed to reverse its direction.

However, as was pointed out earlier, these results alone are not sufficient to establish the existence of active transport. For example, the volume flow of water is not negligible and, therefore, the possibility of a solvent-drag effect cannot be overlooked. Such effects are certainly observed for uncharged molecules such as urea, where the volume flow does produce a net flux of urea opposite to that expected on the basis of the concentration gradient. In this case, however, experimental results seem to indicate that solvent drag can be considered as a minor contributor to the Na-flow. The best argument for this position is that net Na-flow from the mucosa to the serosa is not accompanied by net flow of Cl. The components of the Cl-flux are fairly large, and intestinal tissue, as will be seen, is quite permeable to Cl. Therefore, if the Na-flux was strongly affected by solvent-drag, it would then be very difficult to explain why Cl was not affected, as is clearly the case.

The view that the sodium flux from the mucosa to the serosa is due to active transport is often supported by the fact that inhibition of the metabolic processes in the absorptive cells of the intestinal epithelium is accompanied by reduction in the mucosal to serosal flux. Nevertheless, this is not really certain evidence. Electro-diffusion, for example, is a passive process. If the electric potential is maintained through energy provided by the cell's metabolism, it may be that interference with the metabolism simply eliminates the potential, which will obviously change the flux. In our particular case, measurements of all the relevant quantities are available, and it is possible to eliminate the various uncertainties and conclude that the flux from the serosa to the mucosa is a passive process, while in the opposite direction it is active.

In the case of chlorine ions, Equation 11.1 does seem to predict accurately the flux. Furthermore, *in vitro* experiments show that when Δa and $\Delta \psi$ go to zero, $J_{Cl} \cong 0$. Unidirectional fluxes are also correctly predicted for cases where $\Delta \psi$ is maintained at various values by external means. However, the analysis of *in vivo* chlorines fluxes can be complicated by possible hormonal effects.

Experimental results indicate that the water flow is strongly determined by the chemical potential gradient and, if we introduce either hypertonic or hypotonic solutions into the gut, rapid water inflow or outflow, respectively, is observed. Furthermore, solutions placed in the gut *in vivo* are soon found to be isotonic with the plasma. In addition, no water flow has been observed to occur in the absence of solute flow or, more importantly, in the absence of osmotic or fluid pressure differences. All of this information argues fairly convincingly for the absence of any active transport process that is peculiar to the water flow. However, this does not mean that there are no complications. Although water transport in the absence of solute flow is not observed, there is very good evidence for a connection between water transport and the transport of solutes. Thus, it is found that under certain conditions, the rate of water absorption in the gut is linear with the sodium flux.

This situation has been analyzed by Curran using the model shown schematically in Figure 11.3. As you know, van't Hoff, from an analogy with the kinetic theory of gases, showed that the osmotic pressure for a membrane permeable to solvent but impermeable to solute was given by

$$\Delta \Pi = RT \, \Delta C_i$$

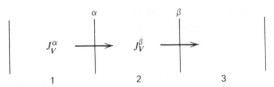

Figure 11.3 The three-compartment model used in Curran's analysis of water flow in the gut.

where ΔC_i is the difference in solute concentration across the membrane. Of course, real membranes are more likely to be semipermeable. Staverman suggested that this situation could be dealt with by introducing a reflection coefficient that measured the permeability. Thus

$$\Delta\Pi = \sigma_i\,RT\,\Delta C_i$$

where $\sigma_i = 1$ if the membrane is completely impermeable and 0 if it is totally permeable.

In Curran's analysis, we assume

$$\sigma_\alpha > \sigma_\beta$$

Therefore, the solute will accumulate in the second chamber and produce an osmotic pressure given by

$$\Delta\Pi_\alpha = \sigma_\alpha\,RT\,\Delta C_\alpha$$

This means that a volume flow of solvent will occur from the first to the second chamber, which will be described by

$$J_{12} = L\sigma_\alpha\,RT\,\Delta C_\alpha$$

where L is the proportionality constant between J and $\Delta\Pi$, often called the fluid or hydraulic conductivity. Since

$$\sigma_\beta < \sigma_\alpha$$

then

$$\Delta\Pi_\beta \ll \Delta\Pi_\alpha$$

and

$$J_{32} \ll J_{12}$$

Therefore, since V_2 is essentially constant, P_2 will increase. Since the membrane β is more permeable than the membrane α, there will be a flux from the second to the third chamber. Hence, since $J_{12} \gg J_{32}$ and J_{23} is not negligible, the effect is a flux from the first to the third chamber, as required by the experimental results.

The importance of this model is that it provides a mechanism for water transport in the absence of an osmotic gradient; this can be verified with artificial systems. Let the second of the membranes be a porous system. It can be shown that for a double membrane,

$$J_v = L(\sigma_\alpha \, RT \, \Delta C_\alpha + \sigma_\beta \, RT \, \Delta C_\beta)$$

Now if

$$\Delta C_\alpha = -\Delta C_\beta$$

then

$$J_r = L(\sigma_\alpha - \sigma_\beta) \, RT \, \Delta C_\alpha$$

So if

$$\sigma_\alpha > \sigma_\beta \qquad \text{and} \qquad \Delta C_\alpha \neq 0$$

then

$$J_v > 0$$

If in our artificial system we use a dialysis membrane for the first membrane, a flow from $1 \rightarrow 2$ will occur if the solute concentration in the second chamber is high. This fulfills the conditions required for the previous result and a net volume flow $J_{1 \rightarrow 3}$ will be observed independent of the value of $(RTC_3\text{-}RTC_1)$, the osmotic pressure difference.

We know that the complex folding of the inner lining of the gut produces a total surface area that is much greater than the surface area one would estimate on the basis of the external dimensions. Does this difference between the outer and inner areas have any significance for the processes we have been discussing? Consider two

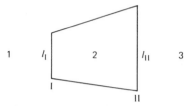

Figure 11.4 The above drawing defines the quantities used to show that transport does not depend on area differences.

membranes with identical properties but different area arranged as in Figure 11.4.

$$J_{\mathrm{I}} = f_{\mathrm{I}}(C_1, C_2) - g_{\mathrm{I}}(C_2, C_1)$$

and

$$J_{\mathrm{II}} = g_{\mathrm{II}}(C_2, C_3) - f_{\mathrm{II}}(C_3, C_2)$$

Then the total flux is

$$F_{\mathrm{I}} = a_{\mathrm{I}} J_{\mathrm{I}}$$

and

$$F_{\mathrm{II}} = a_{\mathrm{II}} J_{\mathrm{II}}$$

In steady state,

$$F_{\mathrm{I}} = F_{\mathrm{II}}$$

and

$$dc_2/dt = 0$$

so

$$a_{\mathrm{I}}[f_{\mathrm{I}}(C_1, C_2) - g_{\mathrm{I}}(C_2, C_1)] = a_{\mathrm{II}}[g_{\mathrm{II}}(C_2, C_3) - f_{\mathrm{II}}(C_3, C_3)]$$

228   **SALT AND WATER TRANSPORT IN THE GUT**

Since the membranes are of identical material,

$$f_I = f_{II} \qquad \text{and} \qquad g_I = g_{II}$$

Now

$$C_1 \doteq C_3$$

and therefore

$$2(a_I + a_{II})(f_I - g_I) = 0$$

We know that

$$a_I + a_{II} \neq 0$$

Consequently

$$f_I - g_I = 0 \qquad f_I = g_I$$

and

$$F_I = F_{II} = 0$$

So there is no net flux merely as a consequence of the difference between inner and outer surface areas.

The problem of sodium transport is further complicated by interactions between the sodium flow and the transport or processing of other substances. For example, if sodium ions are introduced into the gut, glucose uptake by the absorptive cells increases. Furthermore, the reverse is true. If glucose that is actively transported is introduced at the mucosa, the potential difference across the membrane increases as does the sodium transport from the mucosa to the serosa. Such effects are also produced by the addition of other actively transported substances, such as other sugars or amino acids. The membrane can be "fooled" by providing D-glucose, which it cannot use for metabolic processes; D-glucose is still actively transported and the increased potential observed. However, fructose, which can be metabolized but is not actively transported, produces no such effect. This seems clear proof of the central role of the active transport process, whatever its precise mechanism may be. An additional complication in understanding the transport of these substances is that the substance that enters the absorbing cell from the luminal side does not necessarily appear in the same form when it exits to be incorporated into the systemic circulation. For example, fructose reappears as lactate and long-chain fatty acids as triglycerides.

chap. 12

The Behavior Of Striated Muscle

Muscle is a specialized tissue that exhibits contractility to a high degree. Roughly speaking, there are about 500 muscles in the human body. These muscles range from about 1/10 to 70 cm in length and, on the average, account for about half the weight of the body.

BASIC PROPERTIES

There are three types of muscle tissue. The one that will be the major focus of this chapter is the striated or skeletal muscle, which is normally under voluntary control. Such muscle is a composite structure. The macroscopic muscle is composed of fibers; each fiber may be up to 4 or 5 cm long with a diameter of 10 to 100 μm. A fiber is itself a collection of myofibril bundles, with each bundle containing about 10 myofibrils, whose diameter is about $\frac{1}{2}$ μm. The myofibrils are about 20% protein, being composed of actin (MW $\sim$ 450,000; axial ratio 50:1; length $\sim$ 160 nm) and myosin (MW $\sim$ 60,000; axial ratio 12:1; length $\sim$ 30 nm). We note that these two proteins can form a complex, actomyosin, which will shorten when exposed to ATP, and that myosin can act as an enzyme to catalyze the reaction, ATP $\rightarrow$ ADP, which is energy yielding. The myofibril is a gel-like substance. It is also composed of water; salts; a very small amount of two other proteins, tropomyosin and troponin; and ATP. The molecular arrangement is quite orderly, the myofibrial contains thin filaments of actin and thick filaments of myosin. These are arranged in the pattern shown in Figure 12.1. The thick filaments of myosin form the so-called A-band; the I-band is the space between the ends of the bands of the thin actin filaments; H-bands are seen in the A-bands. The dark line in the middle of the I-band is the Z-line. The interval between Z-lines, some 25 μm, is a unit muscle; it is this distance that decreases when contraction occurs.

The second type of muscle tissues, the smooth muscles, are clearly different from the striated. Each fiber from a striated muscle is essentially a unit structure, is connected to one nerve ending, and behaves as a single coordinated structure. In contrast, smooth muscle cells are very much smaller, many cells do not appear to be directly associated with a nerve ending, and the arrangements of the cells is not obviously ordered, nor is the pattern of thick and thin filaments observed. In fact only thin filaments are seen, even though myosin, the material of thick filaments, is easily shown to be present.

Cardiac muscle represents a third type of muscle. Structurally, it shows many features of striated muscle, such as the pattern of thick and thin filaments. However, the cardiac muscle is characterized by the ability of the separate fibers to excite each other, so that the excitation, once initiated, is transmitted from cell to cell. This obviously permits a large mass of cardiac muscle to function as a unit. The contractile process in cardiac muscle is probably very similar to that in striated muscle, but appears more complicated because of the structural complexity of the heart and the ability of the excitation to propagate through the tissue.

The contraction of a muscle is initiated as a charge in the membrane due to the

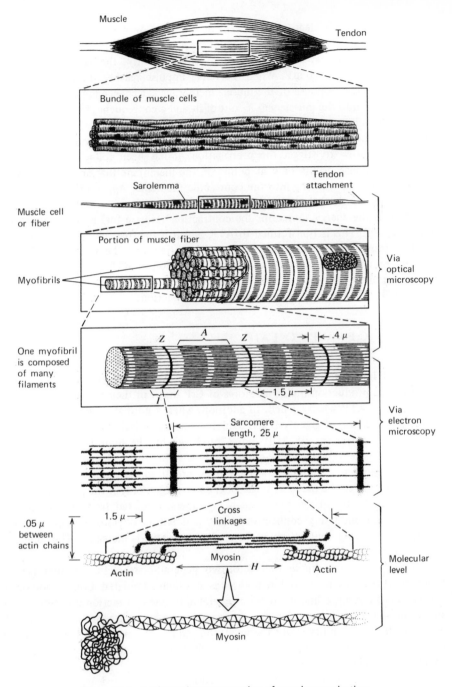

Muscle

Tendon

Bundle of muscle cells

Sarolemma

Tendon attachment

Muscle cell or fiber

Portion of muscle fiber

Myofibrils

Via optical microscopy

One myofibril is composed of many filaments

Z A Z →| |← .4 μ

I |← 1.5 μ →|

Via electron microscopy

Sarcomere length, 25 μ

.05 μ between actin chains

1.5 μ →|

Cross linkages

Actin   Myosin H Actin

Molecular level

Myosin

Figure 12.1 A schematic representation of muscle organization.

235

action of the incoming electrical impulse from a nerve. Consider a striated muscle. The axon to it terminates in a structure known as the end plate, which is specialized for the release of acetylcholine. The muscle membrane (properly, the sacrolemma) has a resting potential of ~ 70 mV. When the nerve impulse reaches the end plate, ACh release occurs. The response of the muscle membrane is a change in permeability with the flow of Na^+ into the muscle and K^+ out of it, and a consequent change in potential. Within a few milliseconds, the cell begins to contract as the local response propagates over the membrane. Furthermore, by studies with microelectrodes operated at potentials too low to trigger the propagation, it has been found that the initial membrane response occurs only at points on the membrane that are marked by the presence of tubules reaching into the contractile filaments. Apparently the excitation changes propagate along these tubules to the interior.

What is the energy source for the contraction? Muscle fuel is stored as creatine phosphate, CrP. This is transformed to ATP, and it is the free energy of hydrolysis of ATP that provides the energy for the muscle contraction. The fact that CrP is the stored energy can be shown by blocking the formation of ATP by any other path. A muscle so treated will still be able to contract until the available CrP can no longer provide ATP. The process is the enzyme mediated reaction:

$$CrP + ADP \rightleftharpoons Cr + ATP$$

In fact, ATP is produced by glycolysis and oxidative phosphorylation when the muscle is at rest; this production builds up stores of CrP that can then be used for the rapid production of ATP when needed. In addition, when a normal muscle is doing work, increases in the rate of glycolysis and oxidative phosphorylation are stimulated by the increased concentrations of ADP and P_i produced by the hydrolysis of ATP. This enables the muscle to replace ATP more rapidly with increasing demand.

In experiments, muscle contraction is produced by applying the required potential directly to the tissue. A single fiber can contract only if the potential exceeds some threshold. In muscle, there are many fibers, insulated from one another by essentially nonconducting connective tissue, and within obvious limits, the muscle undergoes increasing contraction as the applied potential is increased. The time scale for contraction is ~ 100 msec.

Experiments on muscle fall into two general classes. In the first type of experiment, the muscle is fixed at both ends and then stimulated. Contraction occurs without any shortening, and the tension generated by the contraction is measured with an appropriate strain gauge. Contraction occurring under such conditions is said to be isometric. In the second type of experiment, one end of the muscle is free to move and work is done on an attached load. The muscle both contracts and shortens; such a contraction is called isotonic.

Fundamental studies of muscle behavior were pioneered by A. V. Hill. He showed that the maximum tension exerted by a muscle is about 5 kg-m/cm^2 of muscle

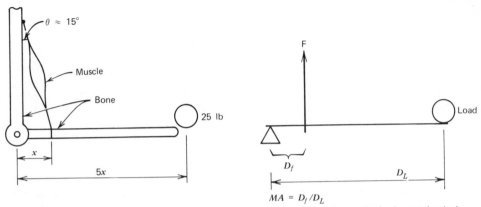

Figure 12.2 The above drawing shows the lever arm arrangement from which the mechanical advantage of the arm is determined.

cross section. This figure may seem a bit large. If we take the circumference of a well-developed upper arm to be 16 in., the circumference of the bicep is about 10 in. This gives a cross-sectional area of about 50 cm^2 and, therefore, a lift capacity with one arm of about 250 kg or about 500 lb. This is obviously not what is observed. We must allow for the fact that the force of muscular contraction is not applied directly, but through a lever action. The sketch in Figure 12.2 indicates the actual situation. The mechanical advantage is

$$MA = \frac{\text{weight}}{\text{force}} = \frac{D_f}{D_1}$$

So

$$F = \frac{25\,(5x)}{x \cos 15°} \doteq 130 \text{ lb}$$

In other words, to support a load as illustrated requires the application of a force that is about five times the load. Thus, in our example, the maximum load with one hand would be about 100 lb, which, given the crudity of our calculation, is roughly correct. The tension is even more extreme if the load is held at the end of an extended arm. The angle between the muscle tendon and the forearm is $\approx 10°$. Then

$$F = \frac{25\,(5x)}{x \sin 10°} \doteq 720 \text{ lb}$$

Hill also showed that in the isotonic case, the tension, P, exerted by the muscle depends on the rate at which the muscle shortens in response to the electrical stimulus.

He found that

$$(P + a)(V + b) = P_0 b$$

where P_0 is the maximum tension, V is the velocity of shortening, and a and b are constants. Although the constants depend on the muscle preparation, the value of P_0/a is roughly constant for all muscle fibers from higher animals.

Hill also studied the heat production by muscle and identified three contributions:

1. The resting heat, due to the metabolic processes associated with living matter, equal to about 2×10^{-4} cal/g.

2. The initial heat, which appears immediately following a stimulus and is produced by the initiation of contraction. The appearance of the initial heat depends on the circumstances of the contraction. In isometric contraction, the initial heat is observed as a sharp rise followed by a more gradual increase. If the contraction is isotonic and the fiber is allowed to decrease in length, additional heat will be generated. The first contribution is called the maintenance heat; the second, the shortening heat. In general, the rate of heat production is linearly dependent on the shortening, but independent of the load. For a single fiber, the maximum temperature increase is about 3×10^{-3} °C, which is equivalent to the generation of $\sim 3 \times 10^{-3}$ cal/g of muscle. If E is the rate of heat production, then

$$E = aV + ab$$

where ab is the maintenance heat. Since the sum of the heat and work is $E + PV$,

$$E + PV = aV + ab + P_0 b - Pb - aV = ab + P_0 b - Pb$$

and therefore the total is linear with the tension.

3. The recovery heat is observed when the muscle returns to rest.

We can obtain a rough estimate of the total efficiency if we divide the sum of the contributions to the heat production into the work done; we find an efficiency of $\sim 20\%$.

MOLECULAR PROCESSES

Szent-Gyorgi discovered that actomyosin threads contract when exposed to ATP. H. E. Huxley, who is not the Huxley of nerve studies, developed a specific model for

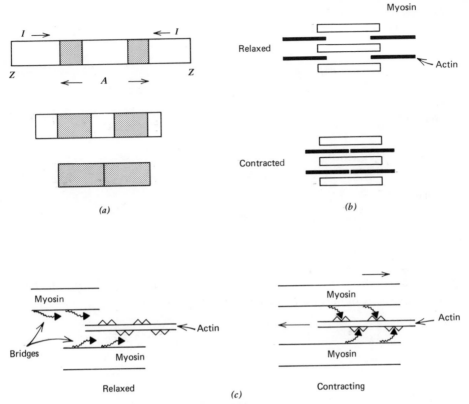

Figure 12.3 A schematic drawing of the sliding-filament model of muscle action, first proposed by Huxley. (a) Microscopic Appearance; (b) Relation of Filaments; (c) Formation of Cross-Bridges The muscle contracts as individual bridges attach to the actin sites, pull the actin strands inward, then release and link to the next site to continue the process. (Of course, all of the cross-bridges aren't formed simultaneously)

the behavior of a muscle fiber. This model takes its cue from the fact that the I-bands are lengthened when a muscle is stretched and shortened when it is contracted, while the A-band remains essentially unchanged. A simple interpretation of these facts (especially after someone else has thought of it!) is that the I-bands slide between A-bands. The A-bands show rather thick structures, and these are the myosin filaments; the I-bands show the thin filaments of actin. Huxley argued that cross-bridges, attached to the myosin filaments, can latch onto sites on the actin filaments. The attachment and "pull" of the filament is accompanied by ATP to ADP conversion, which provides the energy for the motion. A schematic representation of the process is given in Figure 12.3.

The process can be analyzed as follows. Suppose there are two sites: one, M, is on the myosin; the other, A, is on the actin. The formation of a cross-bridge, a "catch," is a process with a rate constant f; breaking the catch is a process with a rate constant g. Let n be the number of links. Then

$$\frac{dn}{dt} = (l - n)f - ng$$

The velocity is

$$V = -dx/dt$$

so we have

$$-V\frac{dn}{dx} = f - (f + g)n$$

How is this velocity related to the actual shortening velocity? If the length of a fiber is s and the velocity of the muscle shortening is v, then

$$V = \frac{sv}{2}$$

Therefore

$$\frac{-sv}{2}\frac{dn}{dx} = f - (f + g)n$$

Let l be the distance between the M-sites. If the work done at one site is

$$\int k x \, dx$$

where k is the Young's modulus of the muscle, it can be shown that

$$P \doteq \frac{msk}{2} \int nx \, dx$$

where m is the number of M-sites/cm^3. By using experimental data to evaluate certain quantities, it is possible to show that this analysis of the dynamics of the sliding-filament model produces results in agreement with Hill's fundamental equation.

240   THE BEHAVIOR OF STRIATED MUSCLE

The mechanochemical interaction between the myosin cross-bridge and the thin actin filaments is the heart of the Huxley sliding filament model. Although the model has been subjected to extensive criticism and review and has undergone a number of improvements, it retains its essential correctness and gives an accurate understanding of the mechanical and thermal behavior of skeletal or striated muscle.

Mittenthal has shown how the model may be extended to cover more general situations. In the Huxley model, the number of cross-bridges that interact with the actin filaments is constant. The cross-bridge that finds itself in a region where it overlaps a portion of the sarcomere can therefore be in one of two states: state C if it exerts a force and state H if it does not. Of course, any particular cross-bridge shifts back and forth from one state to another if the muscle is continuously stimulated. The force exerted by the cross-bridge in a C-state is assumed to be proportional to the distance that the end of the cross-bridge has moved in producing a C-state. Thus

$$F_C = kx$$

where k is a constant, often called the stiffness constant. If $n_C(x, t)$ is the density of the cross-bridges with $x \leqslant x + dx$ at t, the total force is

$$P(t) = \int kx\, n_C dt$$

In terms of our rate constants, $fn_H - gn_C$ is the rate at which n_C changes; n_H is the number of cross-bridges not exerting a force. Let Y be the total number of bridges formed in the overlap region. Now there are two conditions:

$$\int (n_C + n_H)dx = Y$$

which is constant and

$$n_C + n_H = N$$

within a region where attachment can occur. The specific form for the rate constants was assumed by Huxley to be:

1. $f = f_1\, x/h$ in the zone of attachment and 0 out of it.

2. $g = g_1\, x/h$ for $x > 0$ and otherwise a constant.

Table 12.1

Quantity	
v	~5 μm/sec
f_1	~150/sec
g_1	~10^4/sec
h	~4 Å
P_0	~3 kg/cm^2
Y	~5 x 10^{12}/cm^2

Then it can be shown that for isometric contraction,

$$P = P_0 = \frac{f_1}{f_1 + g_1} \, N_0 \, \frac{kh^2}{2}$$

where $N_0 = N$ divided by the fraction of bridges in a region where attachment is possible and h is the length of that region. Characteristic numerical values for these quantities are given in Table 12.1.

The above results are expressly for the isometric case. However, the muscle only does work during isotonic contraction. Roughly speaking, the relation between the tension developed and the particular length at which the muscle is fixed follows the relation shown in Figure 12.4. In the isotonic case, the change in length as a function of load is, within limits, roughly linear. In that case, the maximum work done by the muscle will occur at about half the maximum load. This can be seen by

$$\Delta L = \Delta L_{max}(1 - P/P_0)$$

The work alone is

$$w = P \times \Delta L_{max}(1 - P/P_0)$$

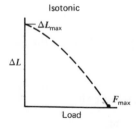
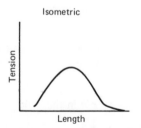

Figure 12.4

and this will be maximum when

$$\frac{dw}{dP} = 0$$

which is when

$$P = \tfrac{1}{2}P_0$$

The dependence of muscle contraction on ATP availability is obvious. Sussman has shown how to find the maximum work that can be made available from the interaction of the contractile filaments and a source of chemical potential. In his analysis, he shows that the conventional expression for the work available from such a system, obtained from

$$\Delta w = \Delta u - T_r \, \Delta S$$

and

$$\Delta w + P_r \, \Delta V = \Delta u + P_r \, \Delta V - T_r \, \Delta S$$

where r indicates the reservoir ensuring that heat exchanges occur at T_r, is not adequate because it does not permit one to distinguish between the work of contraction and the chemical potential exchange. He defines the maximum mechano-chemical work:

$$\int f \, dL \equiv \Delta w_{mc} = \Delta u - T_r \, \Delta S + P_r \, \Delta V - \sum_i \mu_{ri} \, \Delta n_n$$

where f is the force. The availability can then be written

$$\Delta u - T_r \, \Delta S + P_r \, \Delta V - \sum \mu_{ri} \, \Delta n_i - f_r \, \Delta L$$

For constant T_r and P_r

$$\oint f \, dL = - \oint \sum \mu_i \, dn_i$$

Although it appears reasonably clear that the central ideas of the muscle model are correct, it is far from obvious that we have a complete model. One of the most puzzling aspects of muscle behavior comes from Carlson's studies of light scattering spectra from striated muscle. This work began as an investigation into the dynamics of cross-bridge formation.

Muscle, either resting or contracting, carries out two steady-state processes. At rest ATP is slowly hydrolyzed to ADP and O_2 is utilized; this is the process that produces the 2×10^{-4} cal/g observed as the resting heat by Hill. When the muscle is contracted, the same two processes occur, only at a much higher rate. The cycle is shown in Figure 12.5. However, by the application of intensity fluctuation spectroscopy, Carlson has shown that large-scale fluctuations exist that imply a fluctuating driving force, only part of which apparently arises from cross-bridge breakage and formation. Carlson's technique uses a laser beam that is scattered off the muscle tissue target. If the target shows variations in the refractive index on a scale $\approx \lambda$, then the scattered radiation will exhibit periodic diffraction bands. Thus, small-scale structural dynamics may be investigated.

Another problem with the present models has been pointed out by Gray and his objection is potentially quite serious. So far, we have assumed that the formation of a single cross-bridge is a process that can proceed without difficulty because a single ATP molecule can be hydrolized to furnish the energy and be regained by the relaxation of the cross-bridge. Gray has given a strong argument suggesting that the

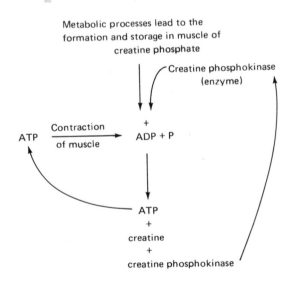

Figure 12.5 A simplified view of the basic energy source for muscular contraction.

explicit requirement of such a process may be in a violation of the uncertainty principle. His argument may be summarized as follows: The cross-bridge must clearly be specified as formed within some small zone Δx. The least energy needed to specify x to within Δx is

$$\Delta E = Hc/4 \, \Delta x$$

The force at x will then be uncertain by

$$\Delta F = F(x) \pm \Delta x(dF/dx)$$

Therefore, over a cycle, clearly we will not come back to the identical situation. Gray proceeds to show that the degree of control required for complete, or near complete, reversibility is much greater than is possible in muscle and, in fact, that degree of control, even for low efficiency, would require some 25 ATP molecules per cycle. From this, he concludes that the true reversible machine with molecular dimensions is not possible.

The question of the control of contraction is also difficult. Galvani's original discovery, in 1791, was that contraction was induced by contact with a metal. That was obviously an electrical effect and we now ask for a molecular mechanism. There is adequate evidence to show that Ca^{++} ions are a controlling factor for the interaction of contractile elements in muscle, and that the passage of these ions is regulated by the sarcoplasmic reticulum. The process proceeds in approximately the following way. The nerve fiber carries a potential impulse to the muscle. This action potential then propagates over the whole length of the muscle fiber with a velocity of about 5 m/sec. The depolarization of the membrane of the reticulum is accompanied by the release of Ca^{++} ions. At least in the case of vertebrate striated muscle, these ions never come from the medium but always from the membrane. The release of Ca^{++} apparently is the critical event that immediately causes the interaction of the thick and thin filaments. As the depolarization dies away, the reticulum membrane begins to accumulate Ca^{++} again, but the Ca^{++} is taken only from the contractile system. As the accumulation progresses, the interaction ceases and the muscle relaxes; hence, relaxation is a passive process. The responses to Ca^{++} by the contractile system occur in all cases investigated thus far; the details of the release of the Ca^{++} and its recapture show some variations. The response of the contractile system to Ca^{++} apparently depends on two proteins found along the fiber and known as troponin and tropomyosin.

chap. 13

Blood Flow And Heart Action

The connection between the physics of fluids, the flow of blood, and the action of the heart is obvious. Indeed, the origin of fluid physics and the study of the flow of viscous fluids began with Poiseuille's investigations of the relation between "the force of the heart" and the "amount of the circulation." Therefore, it would seem that the circulation of the blood, driven by the cardiac pump, would be a topic that could be completely plumbed by recourse to fluid dynamics. Unfortunately, fluid dynamics is a very complex subject ("hydrodynamics is that branch of physics with more paradoxes than laws"), blood is heterogeneous, and the materials of the system are far from elastic. However, many valuable results can be obtained from simple approximations, as well as some preparation provided for those treatments that match the complexity of the subject.

MECHANICS OF FLUIDS

In order to understand the behavior of blood flow, we begin by considering the flow of a perfect fluid. Such a fluid is incompressible and characterized by the fact that when stationary:

1. It exerts an equal pressure in all directions.

2. Any horizontal surface is an isobaryic surface.

3. The pressure is a linear function of depth when in equilibrium with a uniform gravitational field.

These results are the basis of hydrostatics and are often known as Pascal's laws. The third result in the above list is also the physical basis for the measurement of pressures with U-tube and reservoir manometers, in which pressure is measured as the difference in the heights of two columns of fluid supported by pressure. The fluid is often mercury; in that case the pressure that supports a 1-mm column of Hg is

$$\rho g h = 13.6\,(980)(0.1) = 1.3 \times 10^3 \text{ dynes/cm}^2$$

Since $\rho_{blood} = 1.055$,

$$1 \text{ mm of Hg} \equiv 12.9 \text{ cm of blood}$$

Of course, we are not particularly concerned with hydrostatics; we need to consider hydrodynamics. In this case, another way of defining a fluid emerges: A fluid is a substance that cannot withstand a shearing force. The temporary resistance to such a force that may be exhibited by a fluid is due to its viscosity, which may be defined,

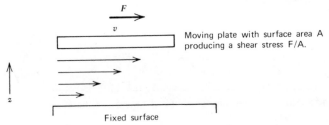

Moving plate with surface area A producing a shear stress F/A.

Fixed surface

Figure 13.1 The quantities used to describe viscosity.

as it originally was by Newton, as "a lack of slipperiness between adjacent layers of fluid." Newton's approach is based on an analysis of the physical situation illustrated in Figure 13.1. He assumed the simplest possible case, namely that

$$\text{the velocity gradient} = \text{the rate of shear} = dv/dz$$

and that

$$\text{the shear stress} = \text{tangential force/cm}^2 \propto dv/dz$$

Then

$$\frac{\text{Shear stress}}{dv/dz} = \text{constant}, \eta$$

which is called the coefficient of viscosity. The unit of η is the poise, the tangential force per unit area with a unit velocity gradient. Thus the dimensions of η are $\text{g/cm}^{-1}\,\text{sec}^{-1}$. The viscosity of water is about 0.01 poise, or a centipoise; the viscosity of normal blood is about 4 centipoise. Of course, it should not be forgotten that water is a so-called Newtonian fluid in that its behavior is accurately represented by the above equation, whereas blood in many cases is not. For that reason, fluids such as blood are called non-Newtonian and are often said to exhibit anomalous viscosity.

One should not confuse frictional and viscous forces. For example, the resistance to blood flow in the arteries and veins has nothing to do with friction. It is entirely the result of the viscosity of the blood.

If we consider a steady flow, in which the velocity at some coordinate is independent of t, and follow the path of a particular volume element in the fluid, we will trace out a streamline. A collection of such streamlines defines a tube of flow. The situation is illustrated in Figure 13.2. The mass of fluid entering one end of such a tube will be

$$dm_1 = \rho A_1 v_1 dt$$

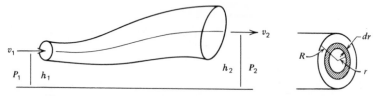

Figure 13.2 The above drawing of a tube of flow shows the quantities used to describe fluid flow.

while the mass exiting will be

$$dm_2 = \rho A_2 v_2 dt$$

where A is the cross-sectional area at the appropriate end of the tube and v is the appropriate velocity. Since we are considering only incompressible fluids, the masses must be the same and

$$\rho A_1 v_1 dt = \rho A_2 v_2 dt$$

Therefore

$$A_1 v_1 = A_2 v_2$$

In the time dt, the mass dm travels a distance equal to vdt. The only force acting on the mass is the pressure (if the ends of the tube of flow are at the same height). Therefore, the pressure must do an amount of work $w_1 - w_2$. Since $P = F/A$, $dw = PAds = PAvdt$. Thus

$$w_1 - w_2 = (m/\rho)(P_1 - P_2)$$

where m is the mass of the fluid that flows through the tube. If there are no frictional forces, this work must produce a change in the kinetic energy:

$$(KE) = (m/2)(v_2{}^2 - v_1{}^2)$$

Since in most cases the ends of the tube of flow are not necessarily at the same level, there is also a potential energy contribution, given by

$$\Delta(PE) - mg\,(\Delta h)$$

Thus

$$\Delta(KE) = \Delta w + \Delta(PE)$$

or

$$P_1 + \tfrac{1}{2}\rho v^2 + \rho gh = \text{constant}$$

This is Bernoulli's equation, the fundamental equation of hydrodynamics. What is the validity of this equation? If we calculate

$$\oint \bar{v} \cdot \overline{dl}$$

we obtain the circulation of the fluid; if the value of the integral is zero, the circulation is said to be irrotational. The physical meaning of this is that the fluid does not possess angular momentum about a given point. If this is the case, Bernoulli's equation is valid throughout the fluid. We should note that the equation is always valid along a streamline. If the above conditions are not satisfied, that is, the fluid is rotational, the general equation is still true, but the value of the constant may change from streamline to streamline.

The above has one important exception: if the streamlines form a set of concentric circles, the circulation is not zero, but Bernoulli's equation still holds throughout the region with the same value of the constant. This can be seen from the following argument.

$$\oint \bar{v} \cdot \overline{dl} = 2\pi r v = \text{constant } K$$

So

$$v = K/2\pi r$$

Since the streamlines are circular, we have a centripetal force:

$$F_C = -mv^2/r$$

and this is balanced by the pressure change across the streamline:

$$F_P = -\Delta P \cdot A$$

where A is the area of the mass element, given by $\rho A\,dr$. Thus

$$\frac{dP}{dr} = \rho\,\frac{v^2}{r}$$

In the limiting case of parallel streamlines, $r \to \infty$, and there is no pressure drop across the flow. In our particular case of circular streamlines,

$$v = K/2\pi r$$

so

$$dP/dr = (\rho/r^3)(K/2\pi)^2$$

or

$$P(r) = -(\rho/2r^2)(K/2\pi)^2 + \text{constant}$$

But

$$1/r^2 = (2\pi v/K)^2$$

whence

$$P + \tfrac{1}{2}\rho v^2 = \text{constant}$$

which is Bernoulli's equation.

Now let us consider the case of viscous flow through a tube of radius R; the physical situation is shown in Figure 13.2a. The fluid is taken to be a series of concentric, cylindrical sheets. By definition the viscous force is given by

$$F_v = \eta A\,dv_x/dy$$

So we have

$$F_v = \eta 2\pi r l\,dv/dr$$

This force is balanced by the pressure difference between the ends of the tube:

$$F_P = \pi r^2 (P_1 - P_2)$$

Thus

$$\frac{dv}{dr} = -\frac{1}{2\eta}\frac{\Delta P}{l} r$$

Now let dQ be the contribution of one of the concentric cylindrical sheets to the total volume flowing through the tube per unit time:

$$dQ = v2\pi r\, dr$$

Then

$$dQ = \frac{\pi}{2\eta}\frac{\Delta P}{l}(R^2 - r^2)r\, dr$$

and

$$Q = \int_0^R dQ = \frac{\pi R^4}{8\eta}\frac{\Delta P}{l}$$

This is Poiseuille's law for a viscous laminar flow.

APPLICATIONS TO BLOOD FLOW

The dependence flow rate on $1/\eta$ has direct physiological consequences. In the condition known as polycythemia, the number of red blood cells/cm^3 of plasma is considerably higher than normal. The interaction between the red blood cells increases the viscosity of the blood. Thus, the flow rate would fall, were it not for physiological responses that act to increase the heart action to maintain normal pressure and flow. The net result is that the heart must increase its work output in order to maintain the circulation. The opposite effect is observed when the number of red blood cells/cm^3 falls below normal, as in the case of anemia. The decrease means that the viscosity of the blood declines, which leads to a higher flow rate since, as before, physiological controls act to maintain normal arterial pressures. Another example of the effect of viscosity is that it increases with decreasing temperature. Therefore, the flow rate in tissues exposed to cold is decreased, with obvious consequences.

Why is blood non-Newtonian? Plasma has a relatively high viscosity ($\sim$ 1.8 cp), apparently due to the presence of asymmetric protein molecules. Since such molecules will take up the most favorable alignment with the flow, we would expect that η for plasma should be velocity dependent, and this is in fact the case. However, plasma is also essentially a Newtonian fluid. It is the addition of the red blood cells that causes the anomalous viscosity. Furthermore, the dependence of η on flow rate is such that at the velocities observed in the major branches of the circulatory system, blood is very nearly a Newtonian fluid.

While the flow through the major arteries and veins is predominantly governed by the Poiseuille equation, the exchange of blood between the arterial and venous systems takes place through the capillary beds, a system of very small tubes whose diameter is approximately the order of a red blood cell. This sort of flow immediately suggests that the Poiseuille equation may be invalid because this situation violates the conditions under which the equation was derived. In fact, the Fahraeus-Lindqvist effect supports that view; the calculated viscosity of blood falls sharply for the case of flow measurements through small diameter tubes.

In order to remove this difficulty, we must replace the integral used to derive the Poiseuille equation by a sum with only a few terms; in the case of such limited flows replacing a sum with an integral is no longer a valid approximation. Thus, if we can get only three cells at a time through a tube, the flow will be proportional to $1^3 + 2^3 + 3^3$, not to $\int r^3 dr$. The difference in the two results is obvious: in the one case we have 36; in the other, $\int r^3 dr = r^4/4 \mid_0^3 = 81/4 \doteq 20$. Hence, if we had used the value from the integral to calculate η, the result would have been seriously in error. Generally,

$$\eta_R = \eta_{PF}/(1 + d/R^2)$$

where d is the diameter of the particle and R is the diameter of the tube. Clearly only the cellular components of the blood are important in this effect since, for all the molecular components,

$$d/R \approx 0$$

and therefore

$$\eta_R = \eta_{PF}$$

Since blood flow does occur with a high density of relatively large particles, it is natural to ask how those particles are distributed in the flow. The distribution of particles, even if they are symmetrical or spherical, will not be uniform across a laminar flow if $dv/dz \neq$ a constant. The Magnus effect, illustrated in Figure 13.3, acts to produce a net force on the particle; this follows from the Bernoulli equation. The direction of this force is toward the direction of highest velocity and lowest gradient in the flow. The net effect is that the particle density in the flow is higher than the overall average if measured at the center of the flow and lower, if measured at the

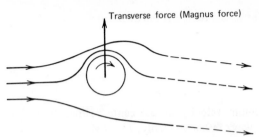

Transverse force (Magnus force)

Figure 13.3 The drawing illustrates the Magnus effect, which is the transverse deflection of a rotating sphere when placed in a uniform flow. In this figure, there is a fluid flow from left to right. In this flow a stationary but rotating sphere is placed. The rotation carries the flow lines around, causing the stream lines at the top of the figure to crowd together. Hence, the top is now characterized by a higher velocity and a lower pressure. The pressure at the top being now lower than at the bottom, the wake of the sphere is turned downward and the sphere moves across the flow, in a transverse direction. The transverse force is maximum when the rotational axis of the sphere is at right angles to the flow. Clearly, as the angle between the rotational axis and the flow goes to zero, the transverse force also goes to zero.

edge. The correctness of this result is confirmed by direct microphotography as well as other theoretical studies of model behavior.

The physiological consequence of this variation in particle density across the flow is that the viscosity of the blood is lower near the walls of the veins and arteries because the red blood cell number density is reduced; conversely, the viscosity is increased near the center. However, an annular ring represents a larger volume near the wall than it does at the center. Therefore, the "wall effect" dominates, and the effective viscosity of the blood in the artery or vein, taken as a whole, is reduced. This reduces the work required to sustain a given flow and the effect is obviously advantageous. The axial concentration of the red blood cells is very sensitive to flow velocity and occurs at fairly low speeds. Of course, this effect is nonexistent in the capillaries where the tube diameter limits the number of cells that can pass. In such cases, the non-Newtonian nature of blood becomes the dominant factor in determining the shape of the pressure-flow curves.

Thus far we have only considered the case of laminar flow in tubes. For Newtonian fluids, the velocity profile across the tube will be parabolic and is given by

$$v(r) = V_m(1 - r^2/R^2)$$

where R is the radius of the tube, r is the distance measured from the center, and

$$V_m = \frac{(\Delta P/\text{cm})R^2}{4}$$

which is the maximum velocity at the center. However, under certain conditions, laminar flow gives way, rather suddenly, to a swirling flow, characterized by eddies. Such a flow is described as being turbulent. The relation between the variables required to describe flow through a pipe is such that turbulent flow will occur when

$$(\rho/\eta)\, vR \geqslant 2 \times 10^3$$

The collection of terms on the left-hand side is often called the Reynolds number. Thus, the critical velocity required to produce a transition from laminar to turbulent flow can be calculated. Velocities that exceed the critical velocity in the case of blood flow may be produced by a variety of causes, some harmless and some not. Laminar blood flow is silent to the stethoscope, but turbulent flow is easily detected. For example, if the access to the ventricle of the heart is reduced, the flow velocity will increase in response to physiological controls and turbulence will occur. Another example is the turbulent flow that occurs at the instant a blood pressure cuff is loosened, which produces the characteristic "tap" heard through the stethoscope.

MECHANICS OF THE HEART

The pressure that acts to drive the blood flow is, of course, generated by the action of the heart. In mammals, the action of the heart (Figure 13.4) may be crudely summarized as a two-step process:

1. The auricular contraction drives the blood into the ventricles; the direction of the flow is controlled by the auriculoventricular valves.

2. The ventricle contraction drives the blood into the aorta and the pulmonary artery. The flow in this step is controlled by the semilunar and pulmonary valves.

It is these two events that produce the characteristic thump-thump of heart action as perceived through the stethoscope.

The work of the heart is supported by the conversion of chemical energy in the mitochondria of the heart muscle. We can estimate the work done by the heart. The

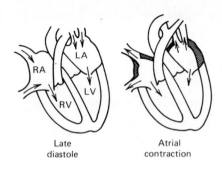

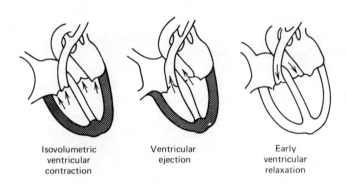

Late diastole Atrial contraction

Isovolumetric ventricular contraction Ventricular ejection Early ventricular relaxation

Figure 13.4 The above figure shows, schematically, the action of the heart. (After Geddes and Baker; App. Biomed. Instru.; John Wiley & Sons; 1968)

kinetic energy of the blood/cm^3 is $\frac{1}{2}\rho v^2$, while the potential is $(\rho gh + p)$. The total energy is

$$H = p + \rho gh + \tfrac{1}{2}\rho v^2 \ \text{erg/ml}$$

Since the volume of the blood is constant, the flow rate, dv/dt, can be taken as constant over an average. Thus H is the amount of work that the heart must do per cm^3 of blood. If the volume of blood moved per contraction is q, the work of the heart is

$$W = qH$$

Suppose we ignore the small height difference between the aorta and the center of mass of the blood in the heart. Then

$$W = q(\tfrac{1}{2}\rho v^2 + p)$$

Since the work done per unit time is the power, π,

$$\pi = \frac{q(\tfrac{1}{2}\rho v^2 + p)_{RS} + q(\tfrac{1}{2}\rho v^2 + p)_{LS}}{\text{time for one cycle}}$$

where the subscripts RS and LS indicate the average values for the right and left sides of the heart, respectively.

The arterial pressure varies because of the heart beat; the maximum pressure is termed systolic; the minimum, diastolic. The pressure in the vein is both smaller and essentially constant. (Hence, the classic sign of a severed artery is a spurting flow, while a severed vein has a steady flow.) Since dv/dt is essentially constant, but $p_A > p_V$, it is possible for the veins and arteries to have similar diameters – ~ 0.5 to 12.5 mm. Structurally the two are very different, with the arterial walls thick but elastic and the walls of the veins thin. The veins have internal valves to prevent incorrect blood flow; the pressure gradient is sufficient for that in the arteries and they have no internal valves.

From physiological experiments we know that

$$v_{LS} \simeq v_{RS}$$

and that the pressure in the aorta is about six times the pressure in the pulmonary artery; therefore,

$$6p_{RS} = p_{LS}$$

Thus

$$\pi = \pi_{\text{hydrostatic}} + \pi_{\text{kinetic}}$$

$$= \frac{7}{6}\bar{p}_{LS}\frac{q}{t} + \bar{v^2}\rho\frac{q}{t}$$

Now

$$\frac{q/t}{\text{Cross-sectional area of aorta}} = \bar{v} = \frac{q/t}{A}$$

and

$$\overline{v_L{}^2} \doteq 3.5(\bar{v})^2$$

so

$$\overline{v_L{}^2} \doteq 3.5q^2/A^2$$

which gives

$$\pi \doteq (\tfrac{7}{6}\bar{p}_{LS} + 3.5\rho)q^3/A^2 t^3$$

For blood, $\rho \approx 1$ g/ml; $A \simeq 4.5$ cm^2. Some particular results are shown in Table 13.1.

The response of the heart to stress can be analyzed in another way. The normal cardiac output is produced at about 70 beats/min for a stroke volume of about 70 ml. The maximum rate for the heart is about 180 beats/min for a stroke volume of about 1/10 liter. This is accompanied by a great increase in the force of the contraction, and therefore the stroke volume enters the artery at a greater pressure and in a shorter time than when the blood is being pumped by the resting heart.

There is a simple way to describe this situation. Let the cardiac output in liters per minute be Q and the amount of oxygen in the venous blood be V cm^3/liter. Then

$$Vq \text{ cm}^3/\text{min}$$

is the rate at which oxygen arrives at the lungs in venous blood. Arterial blood leaves the lungs at q liters/min; its oxygen concentration is C cm^3/liter. Therefore, the rate at which oxygen leaves the lungs is Cq cm^3/min, and this oxygen rate must equal the rate at which oxygen is carried to the lungs by the venous blood and by inhalation. Let the rate at which blood is oxygenated be O_2 cm^3/min. Then

$$Cq = Vq + O_2$$

Table 13.1

	Approximate Values	
	Resting	Exercising
$\bar{p}_{LS}$	100 mmHg	100 mmHg
q	5.5 liters/min	35 liters/min
$\pi_{\text{hydrostatic}}$	1.5 W	10 W
π_{kinetic}	0.02 W	4 W
π	1.5 W	14 W

or

$$q = \frac{O_2}{C - v}$$

C can be measured from a sample of arterial blood and O_2 must be the rate of oxygen consumption. Getting V is a bit more difficult and involves working a catheter into the right ventricle by threading it through the right antecubital vein at the elbow, up the arm, into the superior vena cava, and from there to the ventricle by the tricuspid valve. This procedure is a standard way to obtain the cardiac output, as well as pressure data.

Cardiac output can be controlled in two ways. In one of these the end diastolic volume increases; in the other, it does not. The first case occurs by means of an increase in the force of the cardiac muscle (mycocardium) contraction; the second is connected with the stimulation of sympathetic nerves. The end diastolic volume is measured when the volume of blood in the ventricle is maximum, just before the ventricle starts to contract, or equivalently just at the end of ventricle relaxation (diastolic). The contraction of the ventricle produces the stroke volume and the amount left behind, which is not zero even in normal situations, is called the residual volume. Studies show that if the end diastolic volume increases, the force of the myocardium contraction also increases. This result is known as Starling's law of the heart. Clearly, this means that cardiac output will adjust to match venous return; if venous return is up, the diastolic pressure in this ventricle is up, and thus the end diastolic volume. But an increase in end diastolic volume causes an increase in the force of contraction, which increases the cardiac output.

The second way in which the force of contraction is increased is by the effect hormones such as norepinephorine on the rate of firing of the pacemaker cells, cells with a regular electrical activity which stimulate reaction. Norepinephorine is released by nerve fiber endings near the pacemaker cells. Thus stimulation by the sympathetic nervous system can occur.

Experiments show that under exercise, end diastolic volume is essentially constant as cardiac volume rises. Therefore, the stroke volume increases with the force of contraction by the reduction of the residual volume.

As the results in Table 13.1 show, the work of the heart goes mainly into maintaining the pressure. This is why any difficulty with the arterial circulation that produces a pressure increase puts a strain on the heart. For example, suppose the aorta has its radius reduced by 10% because of fatty deposits. From poiseuille's law, we find that ΔP increases by $\sim 45\%$, which requires a power output increase of essentially the same order from the resting-state heart. The result is not exact because we have ignored the elasticity of the arterial walls.

The hydrostatic pressure must be maintained by the tension of the muscular wall of the heart. We can calculate that tension by appealing to the law of Laplace, which states that the pressure difference across a curved surface under a tension T is given by

$$P = T \left(\frac{1}{R_1} + \frac{1}{R_2} \right)$$

where R_1 and R_2 are, respectively, the radii of curvature in the longitudinal and transverse planes. Thus, for a sphere,

$$R_1 = R_2 = R$$

and so

$$P = 2T/R$$

Since the main work of the heart goes into sustaining the pressure, it follows that the work load of the heart is dominated by the requirement for muscle tension. This helps us to understand the origin of the increased load due to high blood pressure; as the pressure increases, the tension must increase linearly. This is aggravated by the fact that, surprisingly, the efficiency of the heart is rather low. The total energy required is the sum of the tension time and the work of producing the flow, integrated over a cycle. If this is divided into the work done on the blood, the efficiency turns out to be some 15% at best.

Laplace's law can also be applied to the veins, arteries, and capillaries. In the case of a tube, the longitudinal radius of curvature is measured along the axis of the tube and is therefore ∞. The transverse radius of curvature is just the radius of the tube, R. Hence

$$P = T/R$$

Using the appropriate physiological data, we find that the tension in walls increases with the radius. This is why, roughly speaking, the elasticity of the walls also increases: it is a way to maintain the tension without requiring an energy expenditure. The elasticity of the walls is not a Hooke's law force; instead, the resisting force shows a sharply increasing value as the displacement increases.

ELECTRICAL ACTIVITY

As we already know, the activation of muscle is accomplished by the transmission of an electrical signal along the nerve fiber to its connection with the muscle. Thus, it is

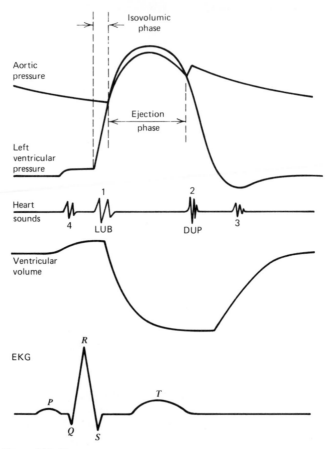

Figure 13.5 The drawing illustrates the cardiac cycle and its relation to the EKG waveform. (Geddes and Baker; App. Biomed. Instru.; John Wiley & Sons)

not surprising that heart activity is characterized by a rather precisely repeated collection of electrical events. These signals have been directly observed since about 1903 by means of electrodes attached to the skin that respond to the electrical potentials generated by the events of the cardiac cycle. The electrodes are attached at a set of arbitrary but agreed upon locations. The normal electrocardiogram, or EKG, may be observed by measuring the potential difference between an electrode attached to the right arm and one attached to the left leg. The maximum potential difference is about 1 mV, and this is usually set to produce a deflection of 1 cm on the EKG. The electrical signal that accompanies a single heart cycle is shown in Figure 13.5, along with the particular heart action associated with the components, or as they are generally called, the waves of the signal.

As a result of practice, a number of standard locations for placing electrodes are accepted; some of these are: left arm, right arm, foot, back and chest. Thus, we could measure the following potential differences:

$$V_1 = V_{LA} - V_{RA}$$
$$V_2 = V_F - V_{RA}$$
$$V_3 = V_F - V_{LA}$$
$$V_4 = V_B - V_C$$

Obviously, $V_2 = V_3 + V_1$, a relation that is often called Einthoven's law. If we take V_1, V_2, and V_3 to be vectors, then the diagram in Figure 13.6, known as Einthoven's triangle, can be constructed. It is assumed that V_1, V_2, and V_3 are chosen so that, taking into account the signs of the potentials,

$$V_{LA} + V_{RA} + V_F = 0$$

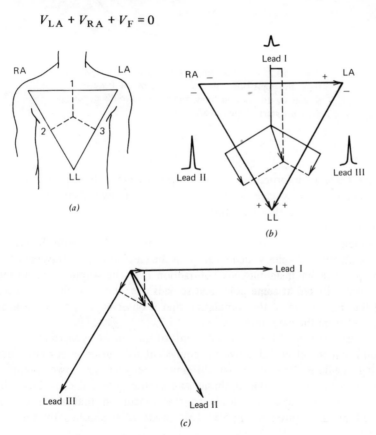

Figure 13.6 Einthoven's triangle.
(Geddes and Baker; App. Biomed. Instru.; John Wiley & Sons)

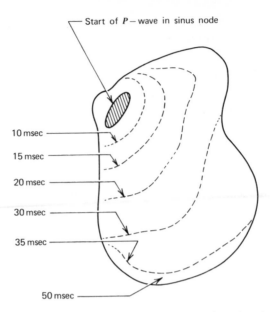

Start of P — wave in sinus node

10 msec
15 msec
20 msec
30 msec
35 msec
50 msec

Figure 13.7 The above shows the right atrium viewed from the right side. The time course of the P-wave is shown by the dotted lines; times are in milliseconds after the start of the wave.

A great deal of information can be extracted from an EKG pattern by someone with an empirical knowledge of the variations that owe their origin to various pathological difficulties. However, this is not a process that is yet explainable in a totally quantitative fashion.

A simple analysis of the origin of the patterns must begin with the fact that the heart is a collection of cells, which means that the electrical signal traversing the nerve does not produce an instantaneous contraction over the whole heart. Instead, the contraction is initiated at some point and spreads over the area, albeit rapidly. Some sense of the time course of the stimulation that is observed as the P-wave in an EKG can be gained from the data shown in Figure 13.7.

A realistic model of the electrical events of the cardiac cycle can be built on the observation that the electrical activity is equivalent to a group of current sources in a conducting medium. The potential difference between any two points in the surrounding tissue is then the algebraic sum of the potential contributed by each current source. This approach leads to the conclusion that the heart may be represented by an equivalent dipole. This result is established by the following argument.

Suppose the position of the ith current source is specified by a vector r_{ij}; this source produces a potential V_i that is a function of the radius vector r. Since there are no free charges,

$$\frac{d^2 V_i}{dr^2} = 0$$

Let the conductivity of the tissue be γ. Then the current density is given by

$$J_i = \gamma \frac{dV_i}{dr} = \gamma \mathbf{E}_i$$

If $V = 0$ at ∞,

$$V(r) = \sum_{i=1}^{n} \frac{I_i}{\gamma \mid r - r_i \mid}$$

$$\simeq \frac{1}{\gamma r} \Sigma I_i + \frac{1}{\gamma r^3} \Sigma (I_i r_i) \cdot \mathbf{r} + \dots$$

Since $\Sigma I_i = 0$, the first term vanishes, leaving

$$V(r) \doteq \frac{1}{\gamma r^3} \Sigma (I_i r_i) \cdot \mathbf{r}$$

But by definition, the dipole moment is

$$\mathbf{p} = \Sigma I_i \cdot \mathbf{r}_i$$

so

$$V(r) \simeq \frac{\mathbf{p} \cdot \mathbf{r}}{\gamma r^3}$$

We can improve the approximation by treating the heart as a sphere; let $\mathbf{p}$ point in the direction $\theta = 0$. Then

$$\mathbf{p} \cdot \mathbf{r} = pr \cos \theta$$

Now

$$V = \frac{1}{\gamma} \, \Sigma \, \frac{I_i}{|\mathbf{r} - \mathbf{r}_i|}$$

but the boundary conditions are

$$V \to \frac{p \cos \theta}{\gamma r^2} \qquad \text{as } r \to 0$$

and

$$\frac{\partial V}{\partial r} \to 0 \qquad \text{as } r \to R$$

which is the radius of the heart. Thus we obtain

$$V(r) = \frac{p \cos \theta}{\gamma} \left(\frac{1}{r^2} + \frac{2r}{R^3} \right)$$

so that

$$V_{r = R} = \frac{3p \cos \theta}{R^2}$$

The above argument shows that we can, in principle, calculate the observed potentials. The argument also shows that the so-called dipole approximation does lead to results that are connected with the properties of the heart. However, it is clear that we cannot, with these simple methods, localize contributions from specific regions of the heart.

Attempts to refine and improve EKG analysis analysis have been underway for some time. One possibility for improvement is to eliminate the assumptions that the heart is in the center of the thorax and that, as our use of spherical coordinates implies, the body is spherical. Thus the number of electrodes can be increased in order to permit one to refine the value of the dipole, or heart vector. Of course, it should be clear that the heart vector changes direction, magnitude, and location over the cardiac cycle, and therefore we are observing a dynamic quantity. If the heart vector can be precisely defined, then the potentials observed at any point on the body's surface are predictable.

Gabor and Nelson have proved this in the following way. Let **r** be the radius vector from the origin to the boundary of a volume dV_{ij}, which is the vector current density, and let s be the source strength. The vector dipole moment is by definition:

$$\mathcal{M} = \int_v \mathbf{r} s \, dV$$

The x-component is given by

$$M(x) = \int xs \, dV = \int x \, \nabla \cdot \mathbf{j} \, dV$$

Then

$$M(x) = \int_v (\nabla \cdot jx) \, dV - \int_v j(\nabla \times dV)$$

whence

$$M(x) = \int_s x \, \mathbf{j}_n \, dS - \int (j \times dV)$$

Now

$$\mathbf{j}_n = 0$$

and

$$j_x = -k \, \partial v / \partial x$$

so

$$M(x) = \int_v k \frac{dv}{dx} \, dV$$

and

$$M(x) = k \int v \, dydz = k \int v ds_x$$

whence

$$\mathcal{M} = k \int v ds$$

where k is the average conductance. Thus, the dipole moment depends on potential measurements on the surface S. Using this approach on a model thorax, measurements predict a value of $M(x) = 0.170$ mA-cm at an angle of $-34.4°$; the actual values were 0.173 and $-35°$.

chap. 14

Physical Aspects Of Hearing

Helmholtz's classic research on the mechanism involved in hearing and vision are well known, and his work on these topics established the investigation of the physical processes of sensory perception as a major topic of biophysics. In this chapter, we will discuss hearing, and in the following chapter we will discuss vision.

A discussion of the physical aspects of hearing obviously requires some knowledge of acoustics and therefore we begin by discussing the elementary properties of sound.

PHYSICS OF SOUND

Sound is the propagation of an elastic disturbance in a continuous medium. Suppose we consider a particle in that medium that is subject to a periodic displacement from its equilibrium position. If ν is the frequency of the periodic displacement ψ, we can write

$$\psi = A \sin 2\pi\nu t + B \cos 2\pi\nu t$$
$$\equiv C e^{j2\pi\nu t}$$

where C is now complex. Then

$$v = d\psi/dt \text{ and } a = d^2\psi/dt^2 = dv/dt$$

These are obviously general equations, and they are valid for any vibrating object.

How is the above related to an actual sound wave? Most sounds are a mixture of frequencies. Fourier showed that any complex tone is equivalent to the sum of a set of pure tones. If the number of pure tones required is finite, we say that the sound wave can be represented by a Fourier series. Suppose the sound wave is $f(\nu)$. Then the sound wave may be expressed as an amplitude function that is continuous with the frequency. This amplitude function, $\phi(p)$, is related to $f(\nu)$ through the Laplace transform:

$$\phi(p) = \int_0^\infty e^{-p\nu} f(\nu)\, d\nu$$

An example of this representation is given in Figure 14.1.

If the velocity of propagation of the sound wave is c, then $\lambda = c/\nu$. In a gas the pressure variations produced by the passage of the sound wave are so rapid that, to a good approximation, the process may be regarded as adiabatic. Since gas cannot resist a shearing force, no transverse wave can propagate in a gas. Therefore, the sound wave must be a longitudinal wave.

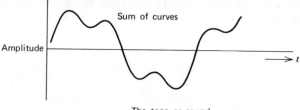

Sum of curves

Amplitude

The tone or sound

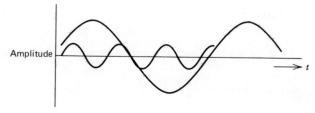

Amplitude

The Fourier series

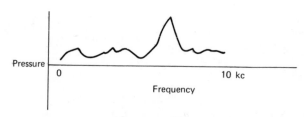

Pressure

0   10 kc

Frequency

The Fourier transform, that is
the spectrum of the sound

Figure 14.1 The above shows three ways to graphically depict sound. In the first, the graph is a direct plot of amplitude versus time; in the second, a Fourier series of several to many terms; in the third, the spectrum, which is the pressure as a function of frequency, is given.

It is known from continuum mechanics that the velocity of a longitudinal wave is given by

$$c = \sqrt{\frac{\text{bulk modulus}}{\text{density}}}$$

The bulk modulus is the ratio of the stress to the strain. We can find an expression for the bulk modulus in a gas by noting that the stress is the force acting on the gas volume per unit area; the strain will be measured by the change in the volume produced by this stress, expressed in terms of the unperturbed volume. Thus

$$\text{Bulk modulus} = \frac{dP}{-dV/V} = -V\frac{dP}{dV}$$

We are assuming the process is adiabatic for a gas. This is not so for a liquid. We can write

$$PV^\gamma = \text{constant}$$

where γ is the ratio of the specific heats at constant pressure and constant volume. So

$$dP/dV = -\gamma P/V$$

and the bulk modulus is given by

$$\gamma P$$

Therefore, in a gas,

$$c = \sqrt{\gamma P/\rho}$$

Now, since $PV = nRT$ and $\rho = Mn/V$ where M is the molecular weight,

$$P/\rho = RT/M$$

This gives

$$c = \sqrt{\gamma RT/M}$$

Thus the velocity of sound in a given gas depends only on temperature, but not on the frequency (or wavelength) of the sound. In a liquid, where our assumptions would no longer be valid, the velocity of sound will depend on the wavelength. This suggests that the acoustical receivers or the acoustical perceptions of an auditory system that has evolved for underwater existence may be quite different from those for airborne sound waves.

Of course, the statement that the velocity is independent of frequency is not absolutely true. When the wavelength of the sound approaches the mean free path of the particles composing the gas, the gas can no longer be regarded as a continuous medium. At STP,

$$mfp = \frac{1}{\sqrt{2}\pi r^2_{molecule}} \doteq 10^{-5} \text{ cm}$$

which corresponds to a limiting frequency of $\sim 10^9$ Hz.

This limit is obviously not of much interest to us. The human ear is sensitive to sound waves in the frequency range between 20 Hz and 20 kHz. However, this does mean that the wavelengths of sound waves are in many cases roughly the same as the physical dimensions of ordinary objects in the environment. For example, at $\nu \approx 35$ Hz, $\lambda \doteq 10$ m; at $\nu \approx 9$ kHz, $\lambda \doteq 3$ cm. When the wavelength of a sound wave is the order of size of a physical object in its path, diffraction effects may be expected. At the low frequencies, diffraction effects will make localization difficult. As the frequency increases, the diffraction effects decline and, at a frequency of several kilohertz, localization is much improved. This situation is reflected by the fact that in animals such as bats that use sound-echo localization, the sound signal is produced at very high frequencies.

The strength of a sound wave is measured by the sound pressure; for a pure tone, such as would be produced by a tuning fork,

$$P = P_0 \sin 2\pi\nu t$$

We usually measure the sound pressure or sound pressure amplitude, p, given by

$$p = P - P_0$$

where P_0 is the equilibrium or ambient pressure. If the sound wave is far enough from the source to be treated as a plane wave, the intensity is expressed in terms of the rms sound pressure $\bar{p}$ and is given by

$$\Gamma \equiv \overline{p^2}/\rho c$$

However, the root-mean-square (rms) pressure amplitude is rather uncommon in actual acoustical discussions of hearing. It is more common to discuss matters in terms of sound pressure levels, defined as

$$L = 20 \log (p/p_0) \text{ dB}$$

where p_0 is a reference pressure that is the threshold of hearing and is usually taken as

$$p_0 \doteq 2 \times 10^{-4} \text{ dynes/cm}^2$$

The above definition can also be applied to the intensity; the intensity level in decibels is given by

$$10 \log \Gamma/\Gamma_0$$

where $\Gamma_0 \doteq 10^{-16}$ W/cm^2. The expressions are not the same because p_0 does not correspond to I_0 except at $T \simeq 300°$K. For $T > 300°$K, the intensity level is always larger than the sound pressure level by $10 \log (T/T_0)^{\frac{1}{2}}$ dB.

The above relations allow us to express the psychological responses to pure tones as physical quantities. For example, from experiment, it is found that, on the average, sound produces pain at a sound pressure level of about 130 dB and detectable physiological damage at about 160 dB. In the latter case, we can find the pressure exerted by the wave and its amplitude by

$$L = 20 \log_{10} (p/p_0)$$
$$160 = 20 \log_{10} (p/0.0002)$$

So

$$p = 2 \times 10^4 \text{ dynes/cm}^2$$

which makes the amplitude of the sound wave about 0.03 atm. Is there an upper limit to L? The answer is yes; if $L \approx 191$ dB, cavitation will occur in the air because of the sound wave, and the air will no longer support the propagation of the wave. The intensity required for this to occur is about 1200 W/cm^2. On the other end of the scale, the ear is not quite able to detect 10^{-20} W/sec. This is about the same as the least energy detectable by the eye. Thus, although it is hard to believe because the sensations are so different, the ear is essentially as sensitive, from the point of view of energy detection, as is the eye.

PHYSICS OF THE EAR

The process of hearing may be regarded as the transduction of mechanical energy in the form of pressure variations in the atmosphere into electrical signals that are processed in the auditory field of the brain. This transduction is shown schematically in Figure 14.2. It begins with vibrations in the ear drum, properly, the tympanic membrane or tympanum, in response to the pressure variation in the atmosphere.

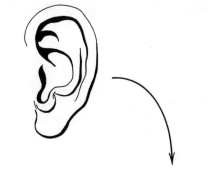

Schematic of physical system

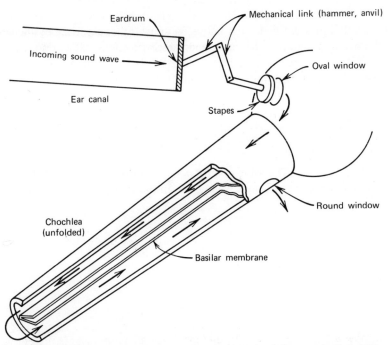

Eardrum

Mechanical link (hammer, anvil)

Incoming sound wave

Oval window

Ear canal

Stapes

Round window

Chochlea (unfolded)

Basilar membrane

Figure 14.2 The structure of the ear and a schematic of the transduction steps.

Motion of this membrane then causes the deflection of a set of small bones in the middle ear. The mechanical linkage of their structure causes a variable compression in the cochlear fluid that follows the atmospheric pressure variation; inward motion of the stapes produces downward motion on the basilar membrane and outward movement of the window membrane. It is the displacement of the basilar membrane that is the "trigger" for the nerve signal. Let us now examine these processes in more detail.

The ear canal, properly the external auditory meatus, is essentially a straight tube about 2.7 ± 0.3 cm long. It is closed at one end by the tympanic membrane. This structure suggests that we should begin by examining the acoustical behavior of a closed pipe, so-called to distinguish it from an open pipe, which is not sealed at either end.

Consider a pipe of length l closed at both ends. Set the air in the pipe to oscillating, say be means of a vibrating disk placed inside the column. The oscillation will be along the pipe, and clearly no motion will occur at the ends of the pipe. The displacement will be $A \sin (\pi X/l) \sin (2\pi t/T)$, where X is the distance along the pipe. It is easy to see that the displacement and the velocity will be maximum at the center of the pipe and be at equilibrium, or at the nodes, at each end of the pipe. On the other hand, the pressure or density change will be zero at the center and maximum at the ends. We have a standing wave with $\lambda = 2l$. This is the fundamental mode of the pipe. Our result would be unchanged if we cut the pipe in half at $X = \frac{1}{2} l$. The midpoint of the pipe is still a pressure node (the only effect now is that air driven by the vibrating disk will move in and out of the pipe), and the closed end is still a pressure antinode.

Now let us consider another pipe of length l. If both ends are open, there must be two antinodes and, therefore, $\lambda = 2l$. Thus the possible frequencies of oscillation, the resonant frequencies, are

$$\nu = nc/2l$$

where $n = 1, 2, 3, 4 \ldots$. The value of ν for $n = 1$ is called the fundamental of the pipe. In the case of the closed pipe, there is only one antinode and so

$$\lambda = 4l$$

Thus the resonant frequencies are

$$\nu = nc/4l$$

However, now $n = 1, 3, 5, 7 \ldots$. In the open pipe all harmonics are present, while in the closed one only the odd harmonics exist. Furthermore, for two pipes of identical size, one open and one closed, the fundamental is at a lower frequency for the closed pipe.

We can now apply the above results to the sound waves that enter our "pipe," the external auditory meatus. First, if the sound can be produced by the vibration of a diaphragm in the pipe, then sound will cause the elastic tympanium at the end of the meatus to vibrate. Since the open end is a node, the pressure on the tympanic membrane is greater than the pressure at the open end of the meatus. The fundamental of the meatus should be about 3.2 kHz, and at this frequency the energy transfer to the tympanic membrane will have its maximum efficiency. Of course, the tympanic membrane is not rigid, but somewhat elastic. Therefore, the resonant frequency is not sharply defined. We can see this by measuring the resonance curve, which is the pressure difference between the ends of the meatus as a function of frequency. We do find the peak approximately as predicted, at about 12 dB for $\bar{\nu} \approx 3.7$ kHz. However, the resonance is broad and can be detected over the range 2 kHz $< \nu <$ 6 kHz. The power output of normal speech is about 200 μW; the power spectrum is maximum for 500 $< \nu <$ 1 kHz and extends from 100 $< \nu <$ 7 kHz. Thus, the resonant range of this relatively low-Q circuit covers the expected frequencies, but peaks about 3 kHz higher than the major speech frequencies.

The tympanic membrane, even at rest, is not a flat disk but rather a shallow bowl with the bottom center pointing in and slightly down. At frequencies of $\nu < \sim$ 2 kHz, the membrane appears to vibrate like a rigid disk restrained by about the upper 30% of the edge. For ν's $>$ 2.5 kHz, that pattern disappears and the tympanum vibrates in segments, in a pattern that depends on the frequencies of the incident wave.

The tympanum forms the outer membrane of the tympanic cavity or middle ear, which is a cavity in the temporal bone and has a volume of about 1 ml. Within this air-filled cavity are the bones (ossicles) that form the mechanical link between the tympanic membrane and the cochlear fluid. On account of their shapes, the bones are known as the hammer, the anvil, and the stirrup, or properly, the malleus, incus, and stapes. These bones serve several functions:

1. They damp any sound conducted by the surrounding bone, as well as any stimuli produced by cranial vibration.

2. They produce amplification of the normally very slight motion of the tympanic membrane.

3. They act as attenuators for any large amplitude waves.

The magnitude of the amplification can be estimated in the following way. The tympanum is about 0.1 mm thick and has an area of about 64 mm^2. The effective area of the tympanium is less than this for $\nu >$ 2 kHz, because only segments of the membrane vibrate; the reduction in area is about 65%. The area of contact between the membrane and the hammer is about 52 mm^2. The mechanical advantage of the ossicles is $\sim$ 1.3. Therefore, the force on the stapes will be given by

$$F_{\text{stapes}} \doteq 1.3\,(52)\,p_{\text{tympanum}}$$

or

$$\frac{F_{\text{stapes}}}{68} = p_{\text{tympanum}}$$

Since the area of the contact between the stapes and the round window is ~ 3.2 mm^2,

$$\frac{F_{stapes}}{3.2} = p_{window}$$

$$\frac{p_{window}}{p_{tympanum}} \doteq \frac{68}{3.2} \doteq 21$$

which is roughly the maximum amplification observed. The average value for humans is probably ~ 17. The difference is presumably due to the effects of friction, which were ignored in the calculation.

The action of the ossicles is such that, with respect to the tympanum, the pressure on the oval window is increased, but the amplitude of the motion is decreased. The net effect is to maximize the transfer of power to the round window and the cochlear fluid, that is from air to liquid. In that sense, the ossicles are an impedance matching device. Of course, the match is not perfect. The impedance is frequency dependent, with a minimum at ~ 1 kHz. This minimum occurs at the frequency where the nonelastic response of the ear, which is large and dominates the low-frequency region, is balanced out by the inertial contribution that derives from the mass of the system and dominates the impedance at high frequencies. These two factors interact to reduce the impedance between about 300 Hz and 3 kHz and to substantially reduce it between about 1 and 2 kHz. Then the system has its most efficient power transfer in the approximate major frequency range for speech.

The point of all the above processes is to get the pressure vibration from the outer tympanum to the liquid-filled inner ear where, in a sense, hearing now occurs. The inner ear is a complex structure; we are concerned only with the fluid-filled cochlea. Within the clchlea, we find the basilar membrane and, in that, a specialized structure, the organ of Corti, which contains the actual nerve endings of the auditory pathway. A large number of physiological experiments have conclusively demonstrated that these two structures are central to the hearing process.

The earliest physical theory of hearing, developed by Helmholtz, argued for the central importance of the basilar membrane in the auditory process. Helmholtz's theory was based on experiments with resonators of his own invention. In their simplest form, these resonators are closed pipes in which the open end has been somewhat constricted; the open end may also be covered by a disk with a hole of area A in it. The common form of the resonator is a cavity from which a short tube is drawn, rather like a Christmas tree ornament. The fundamental frequency of such a resonator will be given by

$$v_0 = \frac{C}{2\pi} \left[\frac{A}{V(t + \delta)} \right]^{1/2}$$

where V is the volume of the cavity, t the neck length, and δ a small correction term given by

$$\delta \approx A$$

The Q's of such resonators are ~ 50.

Using a variety of resonators, Helmholtz analyzed the harmonics or overtones that occurred in various sounds; he concentrated particularly on musical tones because they could be precisely produced by tuning forks. He then attempted to find a component in the ear that would be resonant at the various frequencies that he had shown to be the important components of sound. Anatomical investigations had clearly shown that the basilar membrane was a fibrous structure, and this led Helmholtz to suggest that each fiber was resonant to some particular frequency.

This argument is extremely clever, however, several qualitative difficulties are quickly apparent. In general, the more sharply we define the resonant frequency, the harder it becomes to excite that resonance. Since the ear is in fact very responsive, the tuning of the resonant structure cannot be very sharp. This means that there is a narrow-frequency band, rather than a single frequency, that can excite a particular fiber. However, there are many individuals, especially those with so-called perfect pitch, who can repeatedly distinguish frequencies whose difference is clearly less than the bandwidth required by the model. Another problem is that a resonating system, by its very nature, has an "echo," that is, it continues to vibrate after the excitation of the resonance has stopped; obvious examples of this occur in musical instruments. Such behavior is *not* observed in the hearing process. Sounds are perceived as sharply defined; they cease immediately when the pressure variation ceases. ("Ringing" in the ears is the effect of excessive intensity, not resonance.)

Although the above discussion demonstrates very substantial difficulties with the basilar membrane resonance hypothesis, a strong supporting argument can be advanced. Hearing losses at particular frequencies were often shown by surgical or autopsy results to be associated with damage to particular regions of the membrane. Thus, it could be argued that the hearing loss at those frequencies was due to the loss of response by those fibers resonant at those frequencies.

A more quantitative objection came later, based on the vibrational behavior of the fibers. As can be seen from Figure 14.3, the forces on a stretched string can be written as

$$F_1 = T_0 \sin \theta_1 \simeq T_0 \tan \theta_1 = T_0 (\partial y/\partial x)_A$$

and

$$F_2 = T_0 \sin \theta_2 \simeq T_0 \tan \theta_2 = T_0 (\partial y/\partial x)_B$$

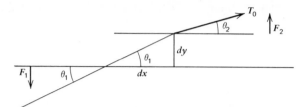

Figure 14.3 The above figure shows the quantities required to analyze the behavior of a stretched string.

where T_0 is the tension. So

$$F_2 - F_1 = T_0 \, \partial^2 y/\partial x^2 \, dx = \rho \, ds(\partial^2 y/\partial t^2)$$

where ρ is the linear mass density, in grams per centimeter. Now $ds \simeq dx$, so

$$\partial^2 y/\partial t^2 = (T_0/\rho)\partial^2 y/\partial x^2$$

This is a wave equation; since we know a stretched string vibrates, we expect an equation of this form. The solution is assumed to have the form:

$$y = \Phi(t)\Psi(y) \qquad \text{so} \qquad \Psi d^2\Phi/dt^2 = (T_0/\rho)d^2\Psi/dx^2$$

whence

$$\frac{1}{\Phi}d^2\Phi/dt^2 = (T_0/\rho)(1/\Psi)d^2\Psi/dx^2$$

Since each side is independent of the other variable, each side must be a constant. Solving each side independently we get

$$\Phi = C \sin \omega t + D \cos \omega t$$

and

$$\Psi = E \sin \omega \sqrt{\rho/T_0} \, x + F \cos \omega \sqrt{\rho/T_0} \, x$$

where the angular frequency ω is given in terms of the length of the string by

$$\omega = \sqrt{T_0/\rho}\, n\pi/l$$

Since $\omega = 2\pi\nu$ and the fundamental is given by the case $n = 1$, the resonant frequency of a stretched string is

$$\nu_{res} = \frac{1}{2l}\sqrt{T_0/\rho}$$

Now, since we know the frequency range over which the ear is sensitive and the lengths of the fibers in the basilar membrane are known from anatomical investigations, we can calculate the range of tensions in the membrane. Measurements by von Békèsey, using techniques that were unavailable to Helmholtz, showed that the tension required by the above equation was at least an order of magnitude greater than the actual tension in the membrane. Consequently, if the resonant fiber model were correct, the audible frequency range would be sharply reduced, by about five octaves.

The mechanism of cochlear response was successfully deduced by von Békèsey, who was able to show that the hydrodynamic characteristic of the interaction between the cochlear fluid and the basilar membrane was the key to the problem. In this work, von Békèsey constructed a number of very ingenious models that were in fact large versions, properly scaled, of the cochlea. With these models he was able to investigate

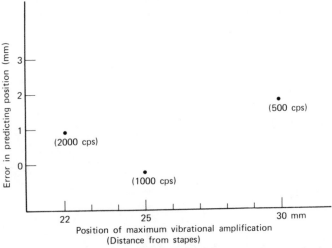

Figure 14.4 von Bekesey's analysis predicts the position of the maximum response on the basilar membrane as a function of frequency. The above graph compares those predictions to the positions determined by direct experiment.

the response of the system to sound and to measure the critical parameters needed to predict the values that should actually occur in the ear. For example, he was able to predict the point of maximum response in the basilar membrane as a function of frequency; the excellent agreement with the actual value is obvious from Figure 14.4.

The hydrodynamics of the cochlea are complex. The process, very much simplified, is that the vibration of the stapes sets up a wave in the cochlear fluid. The wave spreads from the oval window throughout the cochlea in ~ 25 μsec and sets up a traveling wave in the basilar membrane. Since the membrane is less elastic at the base, the displacement moves toward the apex. Thus, the fluid and the membrane form a coupled system, and this is one of the reasons that individual parts of the membrane cannot resonate separately. The position of the maximum displacement is found by taking the envelope of the traveling wave peaks in the membrane at a given instant. This point is dependent on the initial frequency. If the frequency is low the entire membrane moves at the tip; if high, the traveling wave amplitude goes to zero near the base. This is why damage to particular regions causes hearing loss at certain frequencies: high-frequency sensitivity depends on a functional response at the base, while low frequencies depend on the apex being sound. The organ of Corti lays along the membrane and contains the so-called hair cells whose cilia are nearly in contact with the tectorial membrane. Motion of the basilar membrane produces a motion of the cilia, which causes an electrical signal to be generated and then to propagate along the nerve fibers to the auditory cortex of the brain, where it is processed.

Some processing occurs before the signals reach the brain. The threshold potentials required to fire a cell vary and are dependent on the frequency of stimulation; individual nerves are in effect specific frequency channels. Also, as one examines the signals along the path to the auditory cortex, one finds that the firing rate falls. However, the bulk of the processing does appear to occur in the cortex, by mechanisms that so far remain baffling.

chap. 15

Physical Aspects Of Vision

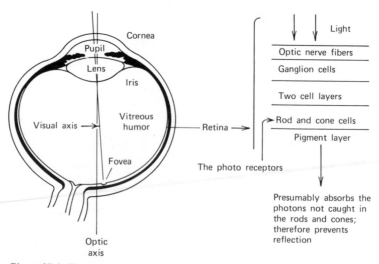

Figure 15.1 The above drawing illustrates, schematically, the structure of the eye and the general organization of the retina. Note that the rods and cones are not at the upper surface of the retina.

Vision is the result of several processes. The physical structure of the eye, shown in Figure 15.1, permits the formation of an image on the retina. The photons that compose this image trigger a series of photochemical events in the light-sensitive receptor cells. This results in the energy carried by the photons being transduced into chemical energy. This chemical energy is transformed to an electrical signal, which is ultimately observed as an electrical impulse propagating along the fibers of the optic nerve to the calcerine fissure in the cerebellar surface of the occipital lobe, where the signals are processed or, if you prefer, translated.

The process of image formation is understood and requires only the application of geometrical optics for its analysis. The transduction of light into chemical energy is fairly well understood. The mechanism for the subsequent generation of the electrical signals, though by no means obscure, is not fully understood; but the details of the mechanism for processing the signals are almost totally unknown.

FORMATION OF THE IMAGE

The image is formed on the retina by the combined refraction of light in the cornea and lens. The refraction of light is described by Snell's law:

$$\frac{\sin \theta_1}{\sin \theta_2} = \frac{n_2}{n_1}$$

where θ is the angle with respect to the normal in each medium, and n is the appropriate value of the index of refraction. In reaching the retina, a light ray undergoes refraction at three locations:

1. Anterior surface of the cornea, in passing from the air into the cornea.
2. Anterior surface of the lens, in passing from the aqueous humor into the lens.
3. Posterior surface of the lens, in passing from the lens to the vitreous humor.

The appropriate indices of refraction are:

$$n_{air} = 1.000$$

$$n_{humors} = 1.333$$

$$n_{lens} = 1.413$$

Thus, the greatest refraction occurs at the posterior surface of the cornea. We can replace the cornea-lens system by an equivalent single lens in order to calculate the optical paths and image sizes. The size of the retinal image is given by the relation:

$$\frac{\text{Object size}}{\text{Image size}} = \frac{\text{object distance}}{\text{image distance}}$$

The distances are measured from the nodal points of the lens. For example, at 1 km an object 40 m high produces a retinal image whose size is given by

$$\text{Image size} = \frac{1}{40}\left(\frac{1000}{0.015}\right) \doteq 0.15 \text{ mm}$$

which is approximately half the size of the fovea. The average image distance is taken as 1.5 cm.

The process of vision is dynamic and the physical process of image formation is no exception. Image distances for various object distances can be calculated from the standard relation:

$$\frac{1}{p} + \frac{1}{q} = (n - 1)\left(\frac{1}{R_1} + \frac{1}{R_2}\right) = \frac{1}{f_L}$$

where

p and q are the object and image distances

n is the index of refraction of the lines

R_1 and R_2 are the front and rear radii of curvature of the lens

f_L is the focal length of the lens

The results of these calculations show that if the focal length of the lens were fixed, then objects near the lens would be focused behind the retina. However, the lens is not rigid and the focal length can be adjusted by changes in the curvature of the lens through a process called accommodation. Helmholtz demonstrated this process in a particularly convincing way. In a darkened room, let a subject relax the eye and "gaze into space." Now hold a shaped light source (Helmholtz used a candle) slightly to one side and observe the subject's eye from the other side. Three images will be seen, one from each of the refractive boundaries. The brightest of the images is produced by light refracted from the cornea surface. A larger and less intense image is produced by the front surface of the lens. The third image is smaller and is inverted; it is produced by light reflected from the rear surface of the lens, which to the observer acts like a concave mirror. If the subject now focuses on a close object, only the second image changes. Its size decreases, and this means that the accommodation is accomplished because the front surface of the lens has reduced its radius of curvature and is now more convex than before. The actual change of curvature is produced by the action of muscles on the tissue in which the lens is mounted. With age, the lens loses its elasticity and thus the power of accommodation declines. This is common with advancing age; those so afflicted typically compensate by holding objects at a greater than normal distance to bring them into proper focus.

RESPONSE OF VISUAL CELLS

The image on the retina is received by the photoreceptor cells known as rods and cones. Each of these cells, shown in Figure 15.2, contains a particular type of light-sensitive molecule called a visual pigment. The spectral absorption curves of these pigments are shown in Figure 15.3. They have been determined from studies of living eyes by measures of the difference spectra between incident light and light reflected from the retina; this method was developed by Rushton and his co-workers. Since the spectral absorption of the pigments is the only component in the total absorption of the eye that will alter upon exposure to intense light, one can be sure that the absorption being measured is really due to the visual pigments by following the change in absorption with time. The association of the particular pigments with the proper receptor cells can also be determined. Normally rods are more abundant than cones by a factor of $\sim 10^2$. However, since in the fovea there are only cones, absorption measures from this part of the retina give only the absorptions of cone pigments. MacNicol's work has clearly shown that the three different types of cone cell are each marked by a single specific visual pigment.

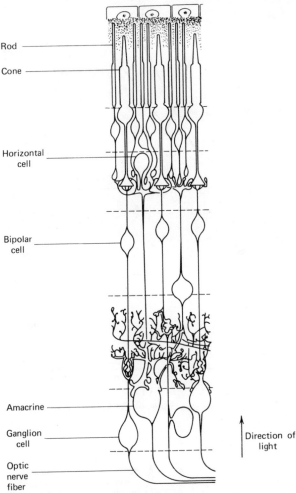

Rod

Cone

Horizontal cell

Bipolar cell

Amacrine

Ganglion cell

Optic nerve fiber

Direction of light

Figure 15.2 A schematic drawing to indicate in somewhat more detail the organization of the rods and cones.

The photosensitive molecule in the vertebrate rods is rhodopsin and it is the best understood of the visual pigments. Rhodopsin has a molecular weight of 35×10^3 Daltons and is composed of a single polypeptide chain, opsin, to which are covalently bonded two molecules, a chromophore, retinal, and a digosacceride. The detailed structure of rhodopsin is uncertain; it resists crystallization and thus no X-ray structure analysis has been possible. From the work of Wald, it is now known that the first step in the energy transduction process is the absorption of a photon by the chromophore. This produces a conformation change. The chromophore dissociates from the opsin by a multistep process known as the bleaching reaction. The quantum

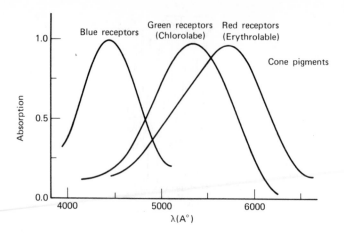

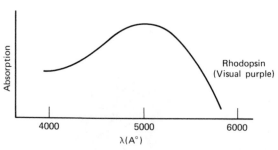

Figure 15.3 The absorption curves of the visual pigments.

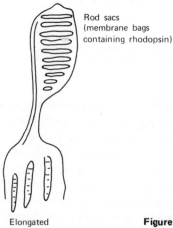

Rod sacs
(membrane bags
containing rhodopsin)

Elongated
mitochondria

Figure 15.4 A schematic drawing
showing the structure of a rod.

efficiency of the bleaching of rhodopsin is ~ 50%. The steps of the bleaching reaction are spectroscopically distinct, but the precise connection between each step and the development of the visual excitation is uncertain. Nevertheless, it is known that the bleaching reaction is mediated by various enzymes and ends with a conformational change in the opsin. In this process it appears that the role of the polypeptide is to direct the conformational change of the chromophore and to increase the quantum yield, relative to the yield that would be produced if the chromophore were free in solution.

Rhodopsin is localized in the outer layers of the rod. About 90% of the area of the outer rod is composed of stacks of several hundred or more disc-shaped membranes. Such a stack structure can be analyzed by low-angle, X-ray diffraction, and the dimensions given in Figure 15.4 are derived from that technique. The rod structure is not static; discs are formed at the base of the outer segment and move, over a period of a few days, to the extremity of the outer segment where they are sloughed off. There is one layer of rhodopsin per disc membrane. The tristructure composed of chromophoric, retinal, and digosacceride layers is apparently more or less embedded in the lipid bilayer of the membrane, although it is possible that the rhodopsin may stick out slightly from the membrane. About 40% (dry weight) of the outer area of the rod is lipid; of the remaining protein, about 80% is rhodopsin. Thus, there are about 60 to 90 lipid molecules per molecule of rhodopsin. The role of the lipid appears to be to protect the stability of the rhodopsin and to facilitate the recombination reaction. On the basis of our current views of membrane structure, we would expect rhodopsin to be very mobile, which seems to be the case.

SENSITIVITY OF RODS

We can assess the sensitivity of rod vision, which is the same as assessing the sensitivity of the retina, by the following experiment, which can be made to yield very precise results. The experiment involves showing a dark-adapted subject a flash of light, with a wavelength and duration that is the most effective for producing a visual sensation. Although particular values obviously vary from individual to individual, the optimal response of the eye appears to occur at $\lambda \cong 510$ nm with a pulse duration of ~ 1 msec. As the intensity of the light flashes declines, some intensity is reached that marks a definite threshold, and any further reduction in intensity produces a decrease in the detectability of the pulses. This experiment was originally performed by Hecht, Shlaer, and Pirenne, who determined the intensity below which less than 60% of the flashes were detected. This intensity corresponds to an energy of ~ 2.1×10^{-10} ergs at the cornea. Since there is then a 4% loss due to reflection from the cornea, another loss of ~ 50% due to absorption in the aqueous and vitreous humors, and a final 20% loss due to absorption in the retinal layers over the light-sensitive cells, it follows that the actual signal at the retina, which produced the visual sensation, was generated, on the average, by about 5 photons. This result can be obtained from the following analysis. Let n be the average number of photons absorbed from a light pulse; n is thus the number of rod excitations produced by the flash. This number must be proportional to N, the average number of photons at the cornea, which is the intensity of the light pulse and is a known quantity in the experiment. Thus

$$n \propto N$$

Since the photon absorptions in the rods are purely random processes, the probability of absorbing m photons can be calculated from the Poisson distribution:

$$P(m) = \frac{e^{-e}n^m}{m!}$$

Let c be the number of events required for the production of a visual sensation. Then the probability of absorbing, at best, c photons will be given by

$$P_c = \sum_{m=c}^{\infty} P(m) = 1 - \sum_{0}^{n-1} P(m)$$

Now the probability of P_c versus n can be calculated for various assumed values of n, which is known to be proportional to N, whose value is known from the experiment. The shape of the plot of P_c versus n depends on c. Therefore, we compare the shape of the experimental curves to the shape of the computed one to determine the correct value of c. Some typical experimental results are shown in Table 15.1.

The major source of the variation in seeing the flash is the variation in the number of photons absorbed in each case, not any fluctuation in the sensitivity of the visual detection mechanism. This can be shown as follows. The area of the light pulse on the retina is $\simeq 0.004$ mm^2. From anatomical studies of the retina, we know this

Table 15.1 Hecht, Shlaer, and Pirene's Results

Average Number of Photons at the Cornea	Frequency of Detection	Prediction for $n = 6$
46.9	0	0.05
73.1	9.4	10.0
113.8	33.3	33.3
177.4	73.5	76.0
276.1	100.0	96.0
421.7	100.0	100.0

means the pulse "covers" about 500 rods. Since there are only some five photons per pulse at the limit of detection of a visual sensation, the chance that a given rod absorbs more than one photon is essentially zero. Therefore, the absorption of a single photon by the rod must be capable of producing the development of a signal. What we are distinguishing with these arguments is the difference between the threshold of the visual sensation (~ 5 photons) and the threshold of receptor activation (~ 1 photon).

If the threshold for the activation of a receptor cell is only a single photon, why does the threshold for a visual sensation require five photons? We can answer this question in the following way.

Studies of the thermal breakdown of the visual pigment rhodopsin in solution indicate that the decomposition is governed by a first-order rate equation and that the products of the breakdown are essentially those of the bleaching reaction. Therefore, we can conclude that rhodopsin cannot tell the difference between thermal and photon energy contributions; in fact, the activation energy, E_A, is found, from the decomposition studies, to be $\sim 44,000$ cal/mole, which is approximately the energy per molecule contributed by a photon with $\lambda = 650$ nm.

We can extend our previous analysis by supposing that the rods also produce signals that are not due to the photons from the pulse, but from another source. Let x be the average number of such events; we can view these events as noise introduced into the detector. Hence, on the average, the total number of events resulting from a light pulse is

$$a = n + x$$

Since the system cannot distinguish between the "noise" in the rod and the absorption of a photon, the number of events required for the production of a visual sensation is still c. The probability of c or more events occurring, if the average number of events is a, is given by

$$P(a|c) = \sum_{n=c}^{\infty} a^n e^{-a}/n!$$

Thus, we have three parameters that are unknown: x, n/N, and c. We also have three parameters fixed by the experiments: the slope of the curve at the threshold, the actual value of the threshold, and the frequency of false responses.

Let us differentiate $P(a \mid c)$ with respect to a; we obtain

$$\frac{dP(a|c)}{da} = \frac{e^{-a}a^{c-1}}{(c-1)!}$$

Since $n \propto N$, we have

$$\frac{dP(a|c)}{d(\log N)} = \frac{dP(a|c)}{d(\log n)} = \frac{dP(a|c)}{da} \frac{da}{d(\log n)}$$

$$= \frac{e^{-a}a^{c-1}}{(c-1)!} \frac{n}{\log e}$$

At the threshold,

$$a = n + x \simeq c$$

By Stirling's formula,

$$c! \doteq \sqrt{2\pi c} \, (c/e)^c$$

Thus

$$\frac{dP(a|c)}{d(\log N)} = \sqrt{2\pi} \log e = \frac{c-x}{\sqrt{c}}$$

When $x = a$, we can use the Poisson equation to calculate the frequency of false responses, $P(x \mid c)$.

Let us explicitly assume that the noise is due to thermal fluctuations. Each rod contains about 10^3 layers of the light-sensitive molecule rhodopsin; each layer contains about 10^3 molecules. There are about 10^8 rods, so the total number of receptor molecules is about 10^{14}. Since we know that activation energy E_A and the molecular weight of the rhodopsin, we can use the Boltzmann factor to estimate the number of thermal events per rod at physiological temperatures ($kT \approx 0.03$ eV). These arguments show that the rate of spontaneous or thermal signals per rod is about one every ten seconds. The original experiment showed that the threshold for the visual sensation was about 5 photons over an area of the retina encompassing about 500 rods. The spontaneous rate in a total of 500 rods is about 50 rods/sec. The longest pulse duration for which the threshold remains unchanged is about 100 msec. Five photons in 10 msec is 50 photons/sec. Therefore, the threshold for the visual sensation is the light input needed so that the photon generated signal is just equal to or greater than the intrinsic noise of the detector. We have given the numbers per rod. Each rod contains about 10^6 rhodopsin molecules. Therefore, a thermally generated signal occurs at the molecular level at a rate no greater than once every 5 minutes.

Light that is incident perpendicular to the rod axis is more strongly absorbed when E is parallel to the disc plane than when it is not; the ratio is about 5:1. If the incident light is parallel to the axis of the rod, no dichroic effect is observed. This suggests that the chromopheres are probably randomly oriented in the plane of the disc.

ELECTRICAL ACTIVITY

The absorption of a photon produces a potential difference across the receptor cell membrane. In vertebrates, the ionic permeability of the membrane decreases, which means the resistance of the cell increases; the reverse is found to be the case for invertebrates. For an unilluminated vertebrate rod, the dark current flows from the inner to the outer segments. The electrical insulation of the discs raises a problem; there are some 10^7 rhodopsin molecules, and it is difficult to understand how the excitation of one of them can, as it apparently does, produce roughly a 1% change in the dark current.

At any rate, the difference in the resting potentials between the inner and outer segments implies a difference in ionic permeability. The sodium permeability of the outer segment is relatively large. Since experiment shows that the elimination of external Na-ions reduces the resting potential to zero, part of the current in the outer segment is a flow of Na-ions. This suggests that if the positive current from the inner segment is due to K-ions, the process may be like that known in other membrane systems.

The fact that electrical events are associated with the visual process does not mean that we can jump to the conclusion that they are produced by the identical process that gives rise to the transmission of the electrical impulse in the axon. In fact, neither receptor cells (rods and cones), nor the nerve cells that are in direct contact with them produce the classic electrical impulse. Svaetichin has shown that the behavior of the retinal cell potentials suggests that they are rather like photocells. If the dark-adapted retina is exposed to a constant light intensity, the observed potential changes to a new but constant value and the size of the change correlates with the intensity. Thus, the initial information at the retina is processed as a voltage amplitude by the receptor cells; both rods and cones hyperpolarize. Although measurements are difficult, the resting potential of the inner segment is between -20 and -40 mV, while when illuminated,

$$\frac{V}{V_{max}} \doteq \frac{I}{I + \sigma}$$

where V_{max} is the saturation or maximum potential observed. The value of σ is the intensity at which $V = \frac{1}{2} V_{max}$. Roughly speaking,

$$I_{max} \simeq 100 \text{ photons/rod}$$

It is the voltage amplitude V that is the trigger for the spike signal; that signal is actually produced in the amacrine and ganglion cells and is then transmitted along the optic nerve to the brain. It is only after the production of that signal that the number of spikes per unit time becomes proportional to the intensity of the stimulus. Measurements on the ganglion cells support the previous given value of the minimum number of photons required to produce a visual sensation. In the cat, a reliably detectable signal from a single ganglion cell is produced when the light intensity at the cornea is about 10 photons. Since absorption in the cat's eye is about 35%, the minimum ganglion signal is produced if ~ 6 photons reach the retina.

COLOR VISION

The above discussion has been an attempt to provide some insights into the generation of electrical signals following the absorption of a photon in a rod cell. However, other cells, containing other visual pigment molecules, also occur. Although the detailed molecules are different, we assume that the processes are roughly the same, and therefore the essential difference between the various cases is just the difference in the absorption properties of the various pigment molecules. This brings us to the problem of color vision. The origin of this problem is very straightforward. On one hand, Newton demonstrated that the spectrum of white light was continuous and thus that there were an infinitely large number of colors. However, the empirical evidence is that a mix of three specific colors is sufficient to reproduce the sensation of any color. From this, Young argued that there must be three different kinds of receptors.

In 1876, Boll noticed that the dark-adapted retina of a frog, normally pink, was bleached by light. The pigment responsible for the color is the rhodopsin in the outer segments of the rods. Rod-vision is color insensitive. Therefore, rhodopsin is not the pigment of the cones. Furthermore, rods are sensitive to light from any angle; cones respond only to central rays. The blue receptors are observed to be sensitive only to central rays; therefore, the blue receptors cannot be rods. Young's view becomes an argument for three types of cones, each with a different pigment and this is in fact what has now been found. We need now only to introduce Rushton's principle: The output of a receptor depends on the number of photons absorbed, but is independent of the particular wavelength. Color is then the brain's processing of the signals arising from the "relative quantum catch in each of the pigments" and, consequently, "there is nothing green or gray but thinking makes it so."

These results also provide the basis for explaining the various forms of what is commonly miscalled "color blindness." Faulty color vision is a sex-linked genetic defect, affecting about 8% of the male population but only about 0.4% of the female population. The defect occurs as the absence of one of the three visual pigments or as the presence of a modified form of one of the pigments. The different types of faulty color vision and the origin of each are shown in Figure 15.5. In order to understand defective color vision, one should note that the spectral absorption curves for the three

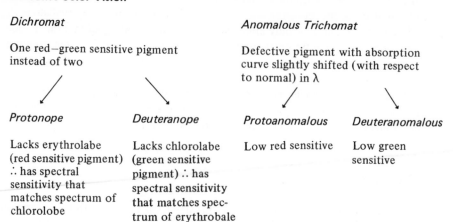

Defective Color Vision

Dichromat

One red–green sensitive pigment instead of two

Anomalous Trichomat

Defective pigment with absorption curve slightly shifted (with respect to normal) in λ

Protonope

Lacks erythrolabe (red sensitive pigment) ∴ has spectral sensitivity that matches spectrum of chlorolobe

Deuteranope

Lacks chlorolabe (green sensitive pigment) ∴ has spectral sensitivity that matches spectrum of erythrobale

Protoanomalous

Low red sensitive

Deuteranomalous

Low green sensitive

Figure 15.5 This chart summarizes the origins of the common color vision defects. Note that the number of dichromats is less than the number of anomalous trichomats.

normal pigments (Figure 15.3) are broad, bell-shaped curves. From these spectral curves, one can calculate the "quantum catch" from incident light of a particular spectral distribution. Because the pigment absorption curves are broad, incident pure red light, for example, produces excitation not only in red-sensitive cones but also in green-sensitive cones. Thus, any defect in either pigment, including its absence, can be expected to produce some kind of effect on the individual's color sensitivity (Figure 15.5).

PROCESSING SIGNALS

How is the information from the receptor cells actually processed? This must begin fairly early in the proceeding. Since the number of retinal receptors exceeds the number of optic nerve fibers by about a factor of 100, some consolidation of pathways, with a resultant consolidation of signals, must begin in the retina. In the frog, optic nerve fibers are found to have different response requirements. Each fiber responds to signals from an area of the retina, and the areas "feeding" different fibers are large enough to overlap each other. This is a puzzling discovery. The lens produces a sharp image, but the signal detection and transmission system apparently acts in such a way as to blur the image. We also find that regions adjacent to illuminated regions are inhibited from producing signals. In the frog retina, about five different types of fibers and ganglion cells are now recognized; for example, there are ganglion cells that act to define boundaries. If there is a boundary in the image, they go "on"; if the illumination is uniform, they do not produce a signal. A more complicated system occurs in the retina of the rabbit, which has some 15 to 18 different ganglion cells. Some of these "fire" only for image movement in a particular direction; some only for images moving within certain speed ranges. These complexities merely serve to suggest the formidable task that must be accomplished in order to understand completely the visual process.

chap.16

On The Origin Of Living Matter

In the discussions in the previous chapters, we have tried to show how physics can be used to understand some of the various properties of living matter. Until now we have avoided dealing with the question of how such matter might arise. In this brief chapter, we now turn to that question, fully realizing that we are now considering a topic in which the pertinent questions are just being formulated.

From the beginnings of chemistry until about 1840, it was firmly believed that organic compounds had their sole origin in connection with the processes of life. Thus, not only did life beget life, it also begat a host of molecules whose properties, behavior, and structure were held to be beyond the realm of "normal" physics and chemistry. By implication, living matter was, therefore, also beyond those laws. Clearly then, the production of organic compounds from totally inorganic sources was impossible, because the particular properties of organic compounds derived from their association with living matter. These arguments constituted the core of the doctrine known as vitalism.

In opposition to the above doctrines was the mechanistic view, which held that the laws of physics and chemistry were applicable to all matter, regardless of how complicated its structure might be. The first major support of the mechanistic view came in 1842 when the German chemist Wöhler succeeded in producing urea, a material of obviously organic origin, from purely inorganic materials. The general trend of developments from that time has been consistently against the vitalistic view. Vitalism, while not disproven in an absolute sense, and still able to count some adherents to a somewhat more subtle form of the doctrine, is now of vanishing significance in any discussion of the origin of living matter.

The mechanistic view suggests not only an understanding of the processes of life but also a "mechanical" origin of life through interactions governed by principles understandable from a physical and chemical viewpoint and in a manner first advocated in detail by Oparin. This attitude has greatly stimulated attempts to delineate the conditions under which life appeared on earth and to suggest the actual mechanisms that would lead to the formation of specific features of living matter. Without doubt, the apparent simplicity of the basic materials has also come to play an important role in the second of the above efforts; only 20 amino acids are needed in order to form proteins with molecular weights of $\sim 500,000$ Daltons and only four bases are needed to make DNA with a molecular weight of $\sim 10^6$ to 10^8 Daltons. Thus the "alphabet" of life appears to be small.

The effort to determine the physical and chemical conditions on the earth at the time life appeared has required contributions from physics, especially astronomy, geology, and chemistry; this effort is a good example of what is needed to deal with an absolutely interdisciplinary problem. The conclusions of the many investigations centered on this problem cannot be treated in detail here; greatly compressed, the results may be summarized as follows. We know that the earth's gravitational field is sufficiently weak so that the average velocity of hydrogen molecules exceeds the escape velocity. We also know from geological evidence, that life appeared in a cellular form fairly early in the earth's history. On the basis of Calvin's work on the chemical "fossils" of life, that is, on material from which the cellular form has vanished, but not the chemical components of that form, the origin may be very early indeed. Therefore, we assume that the rate of H_2 loss from the earth has gone at a constant rate, and we calculate the H_2 pressure in the atmosphere some 4.5×10^9 years ago; the value obtained is 1.5×10^{-3} atm. Chemical arguments show that the pressure due to CO_2 was $\leqslant 10^{-8}$ atm. From arguments based on equilibria considerations, the pressure due to methane, CH_4, is found to be $\sim 4 \times 10^{-3}$ atm. The final central point is the realization that ammonia, NH_3, an easily produced compound, is unstable if provided energy and breaks up into N_2 and H_2. The above analysis was first carried out in detail by Urey, who thus showed that the primitive atmosphere of the earth was a reducing atmosphere composed of H_2, H_2O, CH_4, NH_3, and N_2. The transformation from the reducing atmosphere of the early earth to the present atmosphere with its high-oxygen abundance appears to be a consequence of the action of photosynthesis. From the above, we are forced to conclude that living matter first appeared on the earth in a reducing atmosphere.

The above results provide the cue for the second part of the effort to understand the origin of life: Can we produce the specific compounds known to be essential in living matter under the conditions that have been shown to characterize the primitive atmosphere of the earth? The first experimental success in this direction was attained by Miller and Urey, who showed that an electric discharge in a simulated primitive atmosphere of CH_4, NH_4, H_2O, and H_2 could lead to the production of the amino acids, glycine, alamine, aspartic acid, and glutamic acid. It has also been shown that the amino acids are produced only if the primitive atmospheric mixture is non-oxidizing with respect to methane; the presence of oxides of carbon blocks the formation of amino acids. Of course, if one believes that amino acids were produced in such a way during the early history of the earth, the previous result can be taken as support for the very low predicted abundance of CO_2 in the primitive atmosphere.

Further experiments show that the form in which the energy is provided is not important, and ultraviolet light or ionizing radiation may be substituted for the electric discharge. For example, adenine has been produced by the β-particle irradiation of a mixture of CH_4, NH_4, and H_2O and also by the ultraviolet irradiation of HCN. This latter compound, HCN, is easy to produce under primitive conditions; the UV irradiation treatment of HCN also leads to the formation of guanine. Fox, Ponnamperuma, and others have greatly extended these results, and it is now known that similar processes can produce other critical molecules. For example, adenosine can be produced by the UV irradiation of adenine, ribose, and phosphate.

Indeed, astronomical evidence seems to indicate that the formation of organic compounds is a very natural process. Spectroscopic studies in the radio-frequency range have been particularly successful and have provided conclusive evidence for the presence of some 30-odd organic compounds such as methyl alcohol and formaldehyde, as well as water and ammonia. The radio astronomy results indicate that these compounds are found in the gas and dust clouds distributed throughout our galaxy. Since there is no reason to consider our galaxy as a special case, the same results presumably apply to other galaxies as well. In addition, other significant compounds have been detected by optical spectroscopy; HCN, for example, has been shown to occur in comets. Furthermore, amino acids have been found in the direct chemical analysis of carbonaceous chondrites, a particular and fairly rare type of meteoritic material. The carbonaceous chondrites are soft and rather easily damaged. If not recovered very quickly after they fall and given suitable protection, they are rapidly broken up by weathering. The first organic compounds isolated from such meteorites came from a sample that had been displayed as a museum specimen for some years. This led to considerable discussion about the validity of the results, and it was suggested that the amino acids were contaminants that had accumulated over the years, especially from the specimen being handled. However, similar results have been obtained from a carbonaceous chondrite that was recovered immediately after it fell near Allende, Mexico; there is no doubt about the validity of the presence of original amino acids in this material, which was given complete protection from the time of recovery.

The above results not only show that at least some essential compounds arise through natural processes, but also that such compounds will occur widely. If we believe that the natural formation of these compounds is the first step in the appearance of living matter and that the actual formation of living matter requires only the right conditions, then the above results begin to suggest that living matter may be widely distributed throughout the universe.

Of course, so far only the simplest compounds of living matter have been investigated from the point of view of producing them under primitive conditions. There is a great step between that and producing whole DNA strands or functioning proteins, let alone cellular components or assemblies. However, progress is not entirely absent from these more difficult matters. For example, glycine and leucine tripeptides can be formed by UV irradiation, some polymerization has been observed in the reactants formed by electric discharge in ammonia-methane-water mixtures, and we already know that cell components from different cells of the same type can be assembled to form a functioning cell of that type. Thus, in a certain very crude sense, the possibility of synthesizing a functioning cell has already been proved.

One of the vexing problems involved in understanding the formation of the basic molecules of living matter in the primitive atmosphere is the question of stereo-chemical form. Since the time of Pasteur's work on tartaric acid, it has been known that most organic compounds occur as mirror images of one another. Hence, although the chemical structure is the same, the specific arrangements of the atoms is a mirror image and, consequently, the two molecules cannot be superimposed. For obvious reasons, one type of arrangement is called right-handed and the other left-handed, or properly D and L. In an ordinary synthesis, including the processes that have been used to produce organic compounds under primitive conditions, equal amounts of each form are produced; the mixture is then said to be racemic. Nevertheless, in living matter, this is not the case, and the chemical structure shows an overwhelming preference for compounds in the L-form. The production of a dominant type in the synthesis of a particular compound is by no means beyond the power of the chemists.

In order to accomplish it, however, the reaction must be seeded with reactants whose stereochemical form is that of the one we wish to have predominant in the products of the synthesis. This suggests that living matter is dominated by L-form organic molecules because in the remote past, and perhaps in the very first living matter, that dominance occurred.

A possible clue to the origin of that dominance has been provided by recent work in high-energy physics. As you may know, it was assumed for many years that there was no fundamental asymmetry in natural laws, in the sense that the mirror image of an experiment should give the same result as the original experiment. A different way of expressing this is to say that nature has no preferred direction or orientation. This assumption was called the conservation of parity. Along with the assumption that matter and antimatter particles could be completely exchanged for one another without altering the general outcome of an experiment and that time was symmetric, parity conservation formed a basic law of physics.

Now as far as we know, there are four fundamental forces. These are gravitation, the electromagnetic force, the weak nuclear force, and the strong nuclear force. In the early 1950s, Lee and Yang concluded that in the case of the weak interaction, parity was in fact not conserved. The classic process governed by the weak interaction is β-decay and shortly after Lee and Yang's prediction, Wu and her associates were able to show that the direction of β-particles emitted from aligned Co atoms did indeed have a preferred direction, thereby confirming that parity was not conserved for systems governed by the weak nuclear force.

This may not seem very helpful, since chemical reactions are not controlled by the weak nuclear force but by the electromagnetic force. However, certain experiments in high-energy physics now appear to be best understood by assuming that the weak interaction and the electromagnetic interaction are really just different aspects of the same force. Another way of saying this is that wherever there is an electromagnetic force, there is also a bit of the weak force. Thus, in chemical reactions dominated by the electromagnetic force, there is, in fact, a slight contribution by the weak interaction. This contribution is direction dependent and does "prefer" one form slightly over the other. Hence there arises a slight tendency in favor of L-forms and over some time interval they have come to dominate.

This raises the question of "molecular evolution." Do specific molecular forms suffer alterations over time that lead to changes, perhaps for the better, in parallel to the manner in which evolution appears to occur at a higher level? Although it is clear that changes will occur, it is not obvious that there are any evolutionary consequences to them. Consider the porphyrins, molecules that play an essential role in photosynthesis (in the chlorophyll molecule), in an enzyme critical to the electron-transport process in cellular respiration, and in the oxygen carrying enzyme hemoglobin in higher animals. Porphyrins are formed in experiments carried out under simulated primordial, geochemical conditions. However, it does not follow that porphyrins are somehow the best molecules. In fact, chlorophyll absorption is rather inefficient in just that region of the spectrum where the solar radiation is maximum. Specific arguments by Gaffron suggest that porphyrins are entrenched because they were the first pigment molecule formed that was very stable, at least in comparison to possible competitive pigments. Another way of looking at the specific case of photosynthesis suggests the same conclusion. Molecular oxygen probably made its initial appearance in the atmosphere by the UV photodissociation of H_2O. This produced ozone, which blocked the transmission of UV and therefore stopped further photochemical reactions in the primordal "soup." However, all those systems that had already incorporated a porphyrin were still functional because they could get energy by absorbing light in the visible spectrum. This was the base from which photosynthesis and the resulting high-atmospheric oxygen abundance originated. Thus, porphyrins are predominant because at one point, under very different conditions than now maintain, a porphyrin was the molecular passport to continued existence.

We usually assume that living matter arose from random events. However, we are not saying that a given macromolecule was produced by random assembly. In fact, it is clear that the latter is impossible, as the following argument shows. Suppose we have 100 amino acids and we want to polymerize these into a specific protein. If there are 100 amino acids, there are 10^{130} possible sequences. Thus, even with a reasonably rapid random sampling, the time it would take to construct the specified protein considerably exceeds the age of the universe. However, if some selection process operates, the chance of producing the required protein is greatly increased. Suppose, for example, that once we get an amino acid in the correct position, that position is no longer in hazard. Since there are only 20 amino acids, $N/20$ in any random sequence are correct by chance. Therefore, on the average, it requires $20(N-N/20)$ random tries to get the specified sequence. The value of that expression is $19N$ and, since $N = 100$, we require ~ 1900 tries to assemble the required protein.

Eigen has pointed out that the origin of living matter really involves two questions. The first is whether or not there is a molecule, which would presumably be a protein or a nucleic acid, that can initiate an "organizational" reaction leading to a self-replicating collection of molecules. The second of Eigen's questions raises the possibility that it is the interaction between molecular species, each of which arises independently, that leads to living matter.

There are then two ways the system can develop. Eigen suggests that the system develops to optimize the quality of the macromolecular replication, which presumably proceeds by some sort of template mechanism. However, this mechanism cannot be perfect since, if replication never fails, the system cannot evolve by the selection of favorable random mutations. Therefore, a replication process is required that need be good enough only to insure the continued production of the important macromolecules in a functional form. Prigogine has suggested an alternate evolutionary principle — that the system optimizes the mutations leading to an increase in the entropy production in the system.

Both of these views can be given some analytical formulation. Eigen analyzes his system from the standpoint of information theory; Prigogine prefers the formalism of irreversible thermodynamics. The details of these treatments may be found in the survey papers given in the references.

chap. 17

Bioengineering

Many parts of the previous discussion of biophysics have practical applications, especially through the insights into important physiological processes that such analysis can provide. In addition, during the past several decades, deliberate efforts have been made, often with some success, to develop practical applications by drawing together results from biophysics, other areas of physics, and various branches of engineering. This activity is appropriately known as bioengineering and in this brief chapter some illustrative examples of such work are presented.

NONINVASIVE METHODS: X RAYS

One of the oldest problems is how to observe the internal organs by other than direct surgical procedures. Of course, the ability to observe bone structures was an immediate benefit of Roentgen's discovery of X rays, and such applications were in clinical use within a few months of the discovery. However, simple X-ray techniques have definite limitations. The X rays are a divergent beam that passes through the subject; the images of bones and other tissues are observed as projections on film (Figure 17.1). If tissues overlap, so do images, In addition, X-ray attenuation in soft tissues is low, and tissues that differ only slightly in attenuation are, therefore, difficult to resolve. Attempts to overcome these limitations involved the injection or ingestion of X-ray opaque solutions, such as barium cocktails, or the injection of air into the brain cavities. Although such methods did improve the contrast between different tissues, they were often nontrivial medical procedures, as well as unpleasant experiences for the patient.

These and other difficulties stimulated a search for other noninvasive techniques for studying internal organs. The need for improvement in observational capabilities was emphasized by a growing need for more detailed brain pictures, which were required to match the greatly increased capabilities of neurosurgery.

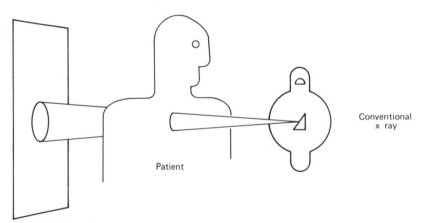

Figure 17.1 A schematic drawing showing that the conventional X-ray observation is essentially a shadow-casting technique.

RADIOISOTOPE IMAGING

A major step forward occurred in 1948, when Moore showed that the injection of certain radioactive materials such as diiodofluorescein was followed by their localization in brain tumors. This meant that the tumors could then be found by scanning the head with a suitable detector.

Anger's invention of a scintillation image camera made the above method extremely useful for diagnostic applications. Anger's camera consists of a large, thin sheet of sodium iodide scintillator. If a gamma ray photon strikes the scintillator crystal, a flash of light is emitted. In Anger's design, a number of photomultiplier tubes that are mounted on the back of the scintillator crystal detect these flashes and register them as current pulses. By placing a collimator over the general region of the tissue being examined, the pattern of the gamma ray distribution and thus the distribution of the radioactive material can be determined.

The success of this approach led to extensive research on other possible applications of nuclear physics to medicine, and four areas were given special emphasis. First, numerous investigations were carried out to determine how radioactive materials could be used to study the functions of various organs. From this came such results as the fact that thyroid activity could be measured by monitoring the uptake of radioactive iodine (I-131). Second, the general techniques for isotope imaging were improved. Third, isotope studies were extended for assaying the properties of tissues obtained by biopsy. Finally, new short-lived isotopes were developed. For example, Tc-99 with a half-life of 6 hours has proved extremely useful because there is no β-decay associated with its radioactivity and, therefore, the radiation dose to which the patient is exposed is sharply reduced, while the γ-ray that is emitted is of low energy, which leads to an increased resolution. Of course, such short-lived isotopes must be produced locally, and this has led to the development of inhouse cyclotron facilities for the production of isotopes such as O-15 ($\tau = 2$ min) or N-13 ($\tau = 10$ min) that are used to measure pulmonary competence.

The net result of these developments was that radioisotope imaging became the method of choice, not only for locating brain tumors but also for studies of liver, pulmonary, bone, thyroid, and kidney abnormalities.

At this point, it is appropriate to note that radiation has played an important role not only as a way of observing the internal organs but also as a therapeutic agent. Radiation therapy was actually in use through radium treatments for skin cancer within a few years after Curie and Curie discovered it in 1898. In 1913 a cervical cancer was reported cured by radium treatments given in 1906.

As you already known from Chapter 10, the proper dose of radiation for therapeutic purposes must be carefully determined and the prescribed dose delivered with an accuracy of $\pm 5\%$ or better. A great deal of work has gone into the design and development of machines that will deliver precise doses, providing more energy to the tumor and less to the surrounding still healthy tissue. For example, relatively low-energy X rays (~ 250 keV) are more strongly absorbed by bone than by other tissues such as muscle. If gamma rays from Co-60 are used, which are much higher energy, bone is about 8% less absorptive than muscle (per gram). Penetration is also important. The maximum dose from Co-60 gamma ray photons is delivered in human tissue at a depth of about 5 mm. For 22-MeV photons produced by a betatron, the depth increases to 4 cm.

DEVELOPMENT OF TOMOGRAPHY

X rays began to regain an important role in brain studies when Hounsfield showed, in 1968, that slit-beam X rays could be used in conjunction with computerized analysis to produce brain scans. This method took its cue from an earlier X-ray technique called tomography. In this method, first used in the 1920s, the X-ray tube moves part way around a circle while the film moves part way in the opposite direction. The patient stands at the center, between the source and the film. Obviously, there is one plane through the patient that contains the axis of rotation of the system. The projected image of this is stationary on the film; everything else moves and produces a generally blurred background. Thus, by careful placement of the patient, adjacent soft tissues can be observed (Figure 17.2).

The mathematical basis for extending this technique was already established by Radon's proof, given in 1917, that the shape of a three-dimensional object could be deduced from a complete set of its projections.

The principles of Hounsfield's design were as follows: The X-ray beam is severely collimated, so that the effective beam is very small. This serves to reduce the scattering and to enhance the resolution. A sodium iodide scintillation system is used to detect the X-ray photons. To study the brain, the patient's head is subjected to a number of exposures, typically 160 equally spaced intervals per each scan direction. Each scan direction is 1° from the previous position. A set of scans is usually taken over a semicircle around the head. The total radiation dose to the skull is about 3 R. The response of the photomultiplier in the scintillation detector will be proportional to the

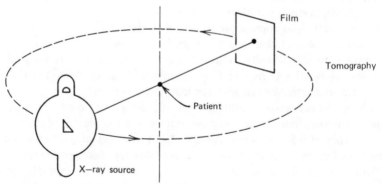

Figure 17.2 A schematic drawing illustrating the basic idea of tomography.

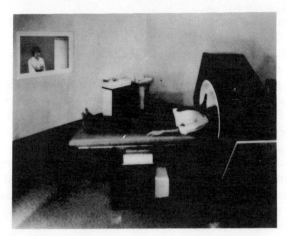

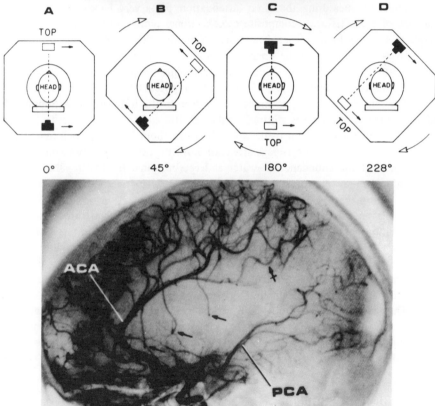

These pictures illustrate the use of computerized tomography for brain examination. The position of the subject in the scanner and the pattern of scanning are shown, as well as the resulting image. ACA and PCA are the anterior and posterior cerebral arteries. The small arrows indicate clinically significant features. (From Gonzalez et al, *Tomography*, Wiley Medical, 1976.)

number of X-ray photons arriving at the detector, and this will depend on the attenuation of the narrow beam by the intervening tissue. The detector response for each of the 28,800 readings is stored in the memory of an on-line computer and analyzed to reconstruct a cross-sectional image of the brain.

The simplest reconstructions are just arrays of the X-ray attenuation values; each number represents the X-ray intensity of one small area of the picture. This small area is often called a pixel. Analysis can be done by an iterative procedure known as ART, for algebraic reconstruction technique. In order to start the iteration, one can, for example, make a reasonable guess about the values for pixels that mark the location of bones with known X-ray attenuation. The program then uses the estimated values to calculate a result and compares it with the real scan. The two will differ and the difference is then distributed over the appropriate pixels for a second run. This is continued until an image of the object, in this case the brain, is obtained that would produce the observed scan results.

Thus, by measuring the X-ray attenuation for a very large number of small volumes of brain tissue, a composite cross-sectional picture of the brain, about 160 pixels square, can be constructed in about 15 minutes. This picture will differentiate between various regions of the brain even though the attenuation coefficients may be only slightly different. Because this method combines the original idea of tomography with a computerized analysis, it is often called computerized tomography or CT.

The sensitivity of the technique can be seen in the following way. Suppose the X-ray attenuation in bone is arbitrarily taken as 100 and that in water as 0. Then the range of attenuation in brain tissue would be from about 2.7 for the white matter to about 4.1 for the gray. The average attenuation for the brain as a whole would be near 2.5 because of the abundance of water and cerebrospinal fluid with values of zero. Conventional X-ray photographs would not distinguish between such materials, but the attenuation coefficients would be correctly measured by CT to about $\pm 1\%$.

The great value of CT is in two areas. First, it can produce three-dimensional representations of internal organs with no discomfort to the patient and second, it permits one to differentiate between adjacent tissues that differ only slightly in X-ray attenuation. However, the pixel size is not particularly small, and therefore the reconstruction does not reveal really fine detail. To reduce the pixel size, one must be willing to greatly increase the computing effort, and consequently the cost.

This means that there is still considerable work to be done toward the development of other noninvasive methods for studying internal structures. One of the most interesting of these is the use of ultrasonic echoes, a method whose clinical potential was first pointed out by Dussik and Dussik in 1947.

ULTRASONIC METHODS

The propogation of ultrasonic waves in the body is a complicated problem. As you know from Chapter 14, sound waves can be represented by several different mathematical expressions. Nevertheless, that earlier discussion was only for purely elastic media in which the energy of the sound wave was always present as mechanical energy, either potential or kinetic. The analysis of ultrasound applications in medicine must begin by recognizing that physiological materials are viscoelastic; that is, when they are deformed, they show viscous and elastic responses simultaneously. This leads to both the dissipation and storage of mechanical energy from the ultrasonic wave, and this means that ultrasonic waves will dissipate heat in physiological materials. In addition, the viscoelastic nature of the material produces a phase lag between the condensation and the excess pressure of the wave.

Since heat is dissipated in physiological materials, the ultrasonic wave of necessity becomes attenuated. The equation for excess pressure now becomes

$$p = P_0 e^{-\alpha z} e^{j(\omega t - \beta z)}$$

where

$$\beta = \omega/c. \qquad \text{and} \qquad \alpha$$

is the absorption coefficient of the medium; α can, of course, be frequency dependent. The intensity can also be written as

$$I = I_0 e^{-2\alpha z}$$

Consequently, the speed of propagation becomes complex and can be written

$$c = \frac{\omega}{\beta - j\alpha} = \frac{\omega}{\sqrt{\beta^2 + \alpha^2}} e^{j\theta}$$

where

$$\theta = \tan^{-1} \alpha/\beta$$

Thus

$$1 = \frac{\omega^2}{\omega^2 + \alpha^2 c^2} e^{2j\theta}$$

From this, it is seen that the simplest way to characterize the ultrasonic properties of physiological materials is to specify the absorption coefficient and the propagation speed. Goldman and Hueter have summarized this data and some general results can be pointed out. Between ultrasonic frequencies of 10^5 and 10^7 Hz, the value of α/f^2, for a variety of tissues, falls off more or less linearly, although the value of α is of course different in different tissues. This is what would be expected on the basis of ultrasonic studies of solutions of macromolecules. The existence of a complex propagation speed implies the operation of a relaxation process. Of course, body tissues are not uniform, and therefore acoustical scattering can also attenuate the beam. In addition, other factors may be important in determining the response of physiological materials to ultrasonic irradiation. For example, the velocity of propagation along muscle fibers is 1592 m/sec, but across them, it is 1576 m/sec. The range in velocities is considerable: in bone, 3360 m/sec; in brain, 1520 m/sec; in fat, 1476 m/sec. Conditions under which exposure to ultrasonic irradiation occurs also must be considered; red blood cell membranes in blood do not rupture if exposed to power levels of $\sim 1\,\mathrm{MW/m^2}$ at 1 MHz. However, if the cell is at a blood/air interface, rupture occurs at powers as low as $0.2\,\mathrm{MW/m^2}$.

These matters are central to the problem of establishing what might be called ultrasonic dosimetry: how to determine the energy transferred to a tissue subjected to ultrasonic irradiation as well as the energy distribution within the tissue at varying times. This is necessary if for no other reason than to assure safety in the medical applications of ultrasound.

Directly connected to the dosimetry problem is the determination of the thermal characteristics of tissues. Heat transfer in materials is characterized by either the thermal conductivity k for steady-state situations or for nonsteady-state situations by the thermal diffusivity

$$\frac{k}{\rho\ (\text{specific heat})}$$

where ρ is the density. Since physiological materials are usually structurally complex, it is usually not possible to calculate the above quantities directly. However, progress is being made in studying the thermal behavior of tissues, and increasingly accurate models can be expected.

Medical applications of ultrasound commonly use echo ranging techniques analogous to those used in conventional sonar systems. The tissue is subjected to either short pulses ($\sim 10\,\mu\text{sec}$; $1\,\text{MHz} < f < 10\,\text{MHz}$; $1\,\mathrm{MW/m^2}$) or a lower-power continuous wave; the first technique is the more common. The ultrasonic waves are reflected at any interface where there is an acoustical impedance difference across the interface. In the case of a plane wave, the acoustic impedance is the ratio of excess pressure to particle velocity, which is ρc where ρ is the density of the medium and c is the propagation velocity. For a tissue with $c = 1500$ m/sec, the time for the wave to travel 1 cm is about 7 μsec. Therefore, if a section of tissue 10 cm thick is to be examined, the maximum time between the initial pulse and the echo is about 140 μsec.

Consequently, the maximum repetition rate is about 7 kHz. The energy that is reflected back, the echo, is displayed as an oscilloscope trace; the echo intensity is a function of the trace brightness. The echogram is then produced by a scan in the direction of propogation; a three-dimensional image is constructed from echograms in different planes. The time required to build up an echogram image is about half a minute. Alternate ways of processing the ultrasonic data are under investigation. Such new methods may lead to acoustical tomography or the medical use of acoustical holograms.

MAGNETIC FIELD MEASUREMENTS

Another noninvasive technique, still very much in the experimental stage, involves measuring the magnetic field pattern of the body, especially near specific organs. These magnetic fields originate in two ways. The first is in response to the ion currents that are constantly traversing the membranes. Thus the current flowing in the cardiac muscle or the brain can be detected both electrically and magnetically. Of course the blood, which is electrically neutral, does not produce a magnetic field. The other way in which magnetic fields may originate in the body is through the presence of foreign, magnetic material.

In any case, the fields are very weak. In the cardiac muscle, the maximum field is 1×10^{-6} G; in the brain, it is 3×10^{-8} G. Because the fields are so weak, special techniques, both for detecting the signals and for isolating the subject from external magnetic field fluctuations are required. These are discussed in the reference by Cohen, one of the pioneers in this work.

THE ARTIFICIAL KIDNEY

Another important area of bioengineering is the design of apparatus that can carry out physiological processes, either for short-term applications or as actual implanted structures. One of the most important of such efforts centers on the development of devices to cope with kidney failure. This effort has two components. One is the effort to understand how the kidney functions, in order to learn what functions are most essential in renal activity, and the other is to develop the apparatus that performs those functions.

A kidney weighs about 300 g. By its action, the volume and composition of the body fluids is held constant over a range of external conditions. The basic structural unit is the nephron, which is composed of a tubule and a glomerulus (Figure 17.3). The nephron is the site of urine production. Renal function is in a sense nothing more than the combined action of the roughly 1.5×10^{6} nephrons in each kidney. Each kidney processes about 0.5 to 0.7 liters/min, which is about 20% of the resting cardiac output. As the blood moves through the kidney, about 20% of the plasma water enters the tubule. Filtration then occurs; some of the substances are reabsorbed into the blood, while others pass into the excretory pathway. For example, all glucose is returned to the circulation, but all *p*-aminohippuric acid will be channeled out as a component of urine. The core of the process is that the flow of the plasma is divided at the glomerulus so that some enters the tubular system while the rest continues in the plasma circulation. The ratio of glomerular filtration to plasma circulation is fixed by the fluid flow equivalent of Ohm's law: The flow is the pressure drop divided by the resistance. The pressure for filtration is then the effective capillary pressure in the tubule, minus any contribution due to osmotic pressure from the proteins in the plasma. Assuming this latter component is described by Van't Hoff's law:

$$\pi = [M]RT$$

we have for the usual protein concentration of 1 mM, an osmotic pressure:

$$\pi = 20 \text{ mm Hg}$$

Since the net capillary pressure is about 50 mm Hg, roughly half the arterial pressure,

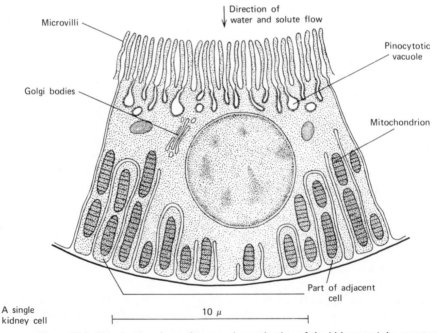

Figure 17.3 This drawing shows the general organization of the kidney and the structure of a kidney cell.

the filtration pressure in the human kidney is about 30 mm Hg, on the average. Denoting this as P_F, the rate of glomerular filtration is

$$K_f P_F$$

where K_f is the ultrafiltration coefficient. (The use of the term ultrafiltration emphasizes that molecular size particles rather than macroscopic particles are being removed.) The permeability of the capillary is then

$$K_f/\text{Surface area}$$

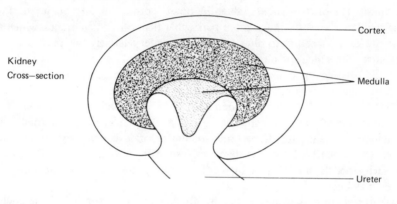

Kidney
Cross—section

Cortex

Medulla

Ureter

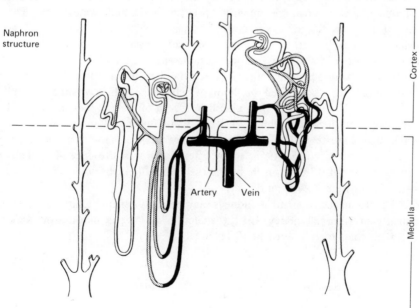

Naphron
structure

Cortex

Medulla

Artery   Vein

THE ARTIFICIAL KIDNEY   319

In rats, for example, $K_f = 4.8$ nl/min^{-1} mm Hg^{-1} and $P_F \simeq 5$ mm Hg. If we assume that that glomerular capillary network can be approximated by a single tube with an equal surface area, then

$$\text{Permeability} = \frac{K_F}{\text{surface area}} = \frac{4.8 \text{ nl/min}^{-1} \text{ mm Hg}^{-1}}{0.0019 \text{ cm}^2}$$
$$= 2.5 \ \mu\text{l/min}^{-1} \text{ mm Hg}^{-1} \text{ cm}^2$$

The permeabilities for other tissues are much smaller than this, ranging from 10% down to about 1% of the above value. In other words, the membranes of the glomerular capillaries are very much more permeable than other membranes, which is reflected in the kidney's ability to filter at high rates under relatively low pressures.

Of course, the kidney is more than a passive filtration device. The membranes exhibit all the complex phenomena we have come to expect, and in addition the system shows involved responses with such matters as the level of the antidiuretic hormone ADH. In spite of the complicated behavior, a very good start on a model for kidney function, capable of predicting such things as the sharply increased urine flow that is found in humans after hypertonic sodium chloride is given by intravenous infusion, has been achieved by a systems analysis due to Bigelow, deHaven, and Shapley. Such models are very valuable in showing how kidney function, or dysfunction, may occur.

Renal dysfunction is characterized by the retention of some 200 metabolic waste products such as urea, creatinine, and uric acid in the body water. Although the identity of the specific toxic substances associated with renal dysfunction were somewhat uncertain, efforts at the removal of such substances from the blood were made as early as the 1920s. These efforts were based on dialysis, the diffusion of the molecules through porous membranes. In 1943, the first clinically useful dialysis units were introduced. They consisted of 20 m of cellophane tubing wrapped around a horizontal drum. The drum was then placed in a bath of dialysis fluid and rotated to carry the blood through the tubing. In 1955, Watschinger and Inouye developed a disposable unit consisting of two cellophane tubes wrapped around a core through which dialysis could occur. In 1965, a new filter design was introduced that replaced the tubing by about 12,000 small, hollow cellulose filters by which the blood is channeled through the dialysis fluid bath.

The performance of artificial kidneys may be analyzed in the same terms as the performance of a normal kidney. Let the volume of blood per minute from which a given solute is completely filtered be V. Then

$$V = \frac{B_F(C_{iB} - C_{oB})}{C_{iB}} \equiv \text{clearance}$$

where B_F is the flood flow in milliliters per minute and C_{iB} and C_{oB} are the input and output concentrations in mg/100 ml. The dialysance D in milliliters per minute is then

$$D = \frac{v(C_{iB} - C_{oB})}{C_{iB} - C_{iD}}$$

where C_{iD} is the concentration of inflowing dialysis fluid. It can be shown that

$$D = V(1 - e^{-kA/v})$$

where k is the membrane permeability and A is the membrane area. The dialysis unit is a mass exchanger that is mathematically describable, just as in the case of heat exchange, by a diffusion equation, specifically Fick's law:

$$J = \mathscr{D}(dc/dx)$$

If $\mathscr{M}$ is the mass transfer rate in moles per minute, then

$$d\mathscr{M} = K_m(C_b - C_d)dA$$

where K_m is the mass transfer coefficient and $C_b - C_d$ is the concentration difference of the solute between the blood and the dialysis fluid. If K_m is constant,

$$\mathscr{M} = K_m A \langle \Delta c \rangle$$

and $1/K_m$ will be the resistance to mass transfer, arising from the resistance of the thin film of blood on the wall of the tubing, the resistance of the membrane, and the resistance of the dialysis fluid. We can write, by analogy with Ohm's law,

$$\frac{1}{K_m} = \frac{1}{K_{blood}} + \frac{1}{K_{tube\ well}} + \frac{1}{K_{dialysis\ fluid}}$$

Using standard fluid-flow equations, it is then possible to calculate from the design of the dialyser the mass transfer rates for given blood flows.

As you know (see Chapter 13), if the tube length is along z, and the pressure is

p, then

$$\frac{dp}{dz} = k = \eta \left(\frac{\partial^2 v_z}{\partial x^2} + \frac{\partial^2 v_z}{\partial y^2} \right)$$

Poiseuille's law gives the total flow per second across a plane perpendicular to the tube:

$$Q = \frac{\pi r^4}{8\eta L} \Delta p$$

where L is the length of the tube and Δp is the pressure drop between the ends.

In order to get the "Ohm's law" form, let the pressure drop be set equal to the resistance times the volume flow. Then

$$\Delta p = R \cdot Q = R \cdot \frac{\pi r^4}{8\eta L} \Delta p$$

so the resistance is

$$R = \frac{8\eta L}{\pi r^4}$$

It is clear that membrane-transport properties play a central role in both the kidney and the artificial kidney. The glomerular membrane will retain serum albumin molecules; the membrane functions as if the pore diameter were about 7.5 nm and the membrane path about 50 nm. For many years, the equivalent artificial filter was cellophane, about 100 μm thick. Now, however, many new membrane materials have come under intensive study. The reason for this is that artificial kidney design was originally aimed at removing molecules such as uric acid and other "waste" molecules whose molecular weight was less than about 300. Analysis of the flow properties of artificial kidneys agreed very well with the observed performance; about 100 ml/min of creatinine with a molecular weight of 113 could be removed by a membrane area of 1 m^2 for a blood flow rate of 200 ml/min. Design studies therefore aimed at reducing the necessary blood input; in other words, at making the process easier on the patient. The rates were about 30 hr/week per patient using about 0.7 m^2 of membrane. However, it was generally realized that artificial kidneys were only partially successful measures and that normal kidneys clearly performed more effectively. For example, the normal kidney removes inuline, with a molecular weight of 5200, at about the same rate as creatinine, with a molecular weight of 113. It is now fairly clear that patients with kidney failure will continue to suffer damage, particularly neurological, unless artificial kidneys can remove solutes up to molecular weights of about 2000. Clearly one way to attain this is through the development of new types of membranes. Some success in this effort is already apparent, and membranes effective up to molecular weights of about 1300 have been made.

New types of membranes have been made from synthetic polymers such as polyacrylonitrile, as well as from substances that might mimic normal kidney membranes such as synthetic polypeptides. Generally such artificial membranes are pictured as impermeable structures containing pores of varying sizes distributed, not necessarily regularly, over the membrane. The pores are, of course, on a molecular scale. The permeability of such a set of pores is

$$D\epsilon/\Delta x$$

where D is the appropriate diffusion constant between the blood and the dialysis fluid, ϵ is the fraction of the surface that contains pores, and Δx is the equilibrium thickness of the membranes. This expression is valid only as long as the size of the solute molecules is less than about 5% of the pore size. Above that, the permeability falls significantly; for example, for solute molecules about 10% of the pore size, the observed permeability is about 80% of the value predicted by the above expression. Of course, if the solute is soluble in the membrane, then the situation is more complicated. In such cases, molecules of approximately equal molecular weight may have very different permeabilities.

Many other problems remain to be completely solved, and it is possible that the solution to the artificial kidney problem will take a completely different form from the one discussed here.

app.

1

Sugges-
tions For
Four Ex-
periments

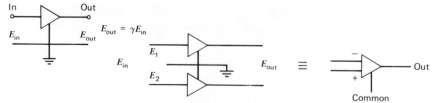

Figure A1.1 A schematic of an operational amplifier circuit.

Many of the signals that are to be measured in the investigation of bioelectric phenomena are small and must be amplified before details in the signal can be observed. You may wish to review the behavior of elementary circuit components given in Appendix 2 before beginning these remarks on amplifiers.

An amplifier can be represented by the circuit shown in Figure A1.1, where ▷ indicates an operational amplifier. The operational amplifier is simply an integrated-circuit device that contains transistors and other components, usually on a small scale. For our purposes the most significant parameter of an amplifier is its voltage gain γ; thus, we write

$$E_{out} = \gamma E_{in}$$

We can also represent the amplifier with a Thevenin equivalent circuit, shown in Figure A1.2, which consists of a source in series with a resistance. Thus, R_{in} and R_{out} are the input and output resistances, respectively. If we want the most effective transfer of the signal from the source to the load R_L, which could be a measuring device, we must match the load to the source resistance. This is clearly the case, since the power dissipated in the load resistance is equal to the voltage drop across the resistance multiplied by the current through it:

$$P = E_L I_L = I_L{}^2 R_L = \gamma E_{in} \frac{R_L}{R_L + R_{out}}$$

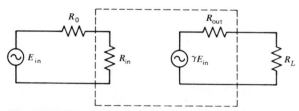

Figure A1.2 The Thevenin equivalent circuit for an operational amplifier.

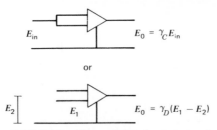

Figure A1.3 The above shows how two operational amplifiers are combined to produce a differential amplifier, one whose output is the amplified difference between two input signals.

and therefore, P is maximum when $R_L = R_{out}$. Impedance matches to within a few percent are usually sufficient.

Although the previous circuit had one input terminal at ground, this is really not very satisfactory and can be avoided by combining two amplifiers (Figure A1.3). If the two amplifiers are balanced properly, the output will be proportional to $E_1 - E_2$; for that reason, this arrangement is called a differential amplifier. Clearly, the relation of the individual potentials to ground is now not important. If there is a large component common to both inputs, then $E_1 - E_2$ is very small. Usually we do not want this component to be amplified very much; our interest is in the difference potential. The amplifier has two operational modes, as shown in Figure A1.3. The value of γ_D/γ_C is a measure of the quality of the amplifier and is called the common-mode rejection ratio (CMRR). The CMRRs of amplifiers used to study bioelectric phenomena should be $\sim 10^4$ to 10^5. Notice that the amplifiers in our schematics were presented for educational purposes and would have very poor CMRRs. The CMRR is usually substantially increased by a feedback from the common mode to the input.

The amplifier must obviously be supplied with an operating voltage; operational amplifiers require low-voltage direct current that is provided from an appropriate transformer. We do not want this operating voltage to be superimposed on the input or output signal. The input and output of the amplifier contain a capacitor, so that no dc level can pass. In addition, the capacitor acts as a filter to reduce noise. An amplifier circuit will therefore contain both capacitors for low-pass filters and resistors for high-pass filters. Conventionally, the low-pass filter is described by the time constant of the resistance-capacitance (RC) combination, while the high-pass is characterized by the frequency limit above which a signal can be transmitted. The high-pass filter will act to cut out low frequencies and, since one of the principal components of noise varies as $1/f$, this is an effective noise reducer. The other usual component of noise, known as Johnson noise, comes from thermal fluctuations and depends on the bandwidth, that is, on the frequency range that is being passed. The high-pass filter has an essentially infinite bandwidth. By combining this with a low-pass filter, the bandwidth can be sharply reduced and thus, the contribution from Johnson noise decreased.

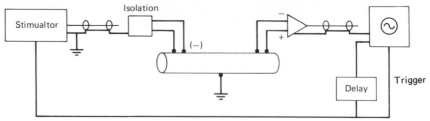

Figure A1.4 A schematic of a general arrangement for studying nerve behavior.

Of course, nothing is free and the cost of filtering the signal to reduce the noise is a slight (or sometimes not so slight) distortion of the signal shape or waveform; rather severe distortion can be produced by simple filters. Such distortion is eliminated by good design and by matching the equipment to the measurement requirements.

The output of the amplifier is normally studied on an oscilloscope, which is really just a very sensitive voltmeter with the ability to follow signal changes that occur in short time intervals. The trace on the scope face is thus just potential as a function of time.

Some of the experiments that follow require the preparation of living material. It is only common sense to proceed with the necessary dissection *after* one clearly understands the procedure and can carry it out clearly and without waste. If you have no experience in these matters, consult someone who has *before* you begin.

The first experiments consist of observations of bioelectric phenomena, particularly the nerve signal. The experimental arrangement for studying the electrical behavior of a nerve fiber is shown in Figure A1.4. The potential that stimulates the nerve impulse is applied to the nerve fiber by a pair of electrodes; a second pair is used to detect the signal produced in response to the stimulus. However, the nerve fiber must be kept moist. The solutions used for this purpose contain ions. This leads to the difficulty of electrode polarization. Ions from the fluid used to maintain the nerve fiber are attracted to the electrodes and form a layer around them, and this interferes with the detection of the signal. Ideally, electrodes should be nonpolarizable and made from Ag/AgCl or cotton wick. However, in the experiments outlined here, the stimulation and signal currents are mostly of very short duration, and electrodes of Ag or Pt wire are not likely to produce particular polarization problems.

The electrodes should be in direct contact with the nerve fiber. The nerve tissue, however, must be moist if it is to retain its ability to function. If the tissue is flooded, or even very moist, direct contact will be difficult to attain. The usual bathing solutions have a low resistance. Since we are trying to measure small potentials while drawing only a negligible current, the impedance of the amplifier must be quite high. Therefore, if there is a liquid low-resistance path, the signal will follow it and the measurements will suffer accordingly.

The necessary amplification can be done in two steps: first in a preamplifier and then in the amplifier of the oscilloscope. All amplifiers should be ac coupled, that is, with a capacitive input. The amplifier obviously cannot tell the difference between the signal you are interested in and any other potential. Therefore, these unwanted signals must be eliminated. The electrically noisy environment is made obvious by just grounding the input lead of the oscilloscope. The other lead then responds to any potential above ground, and the noise from such things as 60-cycle pickup is easily seen on the oscilloscope. In order to reduce the noise, a metal wire cage may be placed around the chamber that contains the nerve; this Faraday cage will shield the electrodes from external electrical fields.

The stimulator pulses have amplitudes of ~ 10 V; the signals are $\sim$ millivolts. Since the preparation is moist, there is a shunt path to the amplifier for the stimulus voltage. The portion of the stimulus voltage that is amplified and observed on the oscilloscope is known as the stimulus artifact. The artifact does not necessarily cover up the signal, whose velocity is less. However, the amplifier's response to the artifact may lead to a distortion of the signal. Therefore, the stimulus artifact should be reduced as much as possible. Aids in this direction are:

1. Do not ground the stimulus electrodes.

2. Use a differential amplifier with a high common mode to avoid contact through ground, which is the "cause" of the artifact.

3. Keep the stimulating electrodes close together

4. Ground an intermediate electrode for the leakage current, and keep this electrode close to the stimulating electrodes.

5. Make sure the nerve fiber is away from the walls of the nerve chamber; fluid film on the walls is a shunt path.

The stimulus artifact can be observed by using a thread that has been soaked in a KCl solution as a substitute for the nerve fiber. Use a 5-V stimulus pulse with a duration of ~ 1 msec. With the oscilloscope sweep rate at ~ 5 to 10 msec/cm, and a sensitivity of 2 to 10 V/cm (or 0.2 to 1.0 V/cm and a preamp gain of $10x$), the artifact should be seen. Test for the artifact by disconnecting the intermediate electrode from ground.

Some bioelectric phenomena are very easy to detect. For example, using graphite or copper electrodes on opposite sides of a bowl of water, potentials are easily observed by just putting your hand in the water between the electrodes and alternately clenching and relaxing the fist. The EKG of a rat is also easy to detect. Anesthetize the rat and, using sterile hypodermic needles for electrodes, insert them just under the skin on either side of the ribs.

The action potential can be observed in an earthworm. Place a large earthworm in 5% alcohol until still; then remove and open the midregion. The nerve cord can be cut away from the surroundings. Place one electrode on the nerve, the other on tissue.

Ringer's solution and mineral oil can be used to prevent the cord from drying. The time base on the oscilloscope should be ~ 10 msec/cm, and the vertical sensitivity about 300 μV/cm. Action potentials can be observed upon mechanical stimulation of the anterior of the worm; tapping with a glass stirring rod is satisfactory.

The EKG of a frog can be observed by pithing a frog and opening it to expose the heart, which should be kept moist with Ringer's solution. Using a slow sweep rate, synchronize the oscilloscope to the heart rate; the EKG can be observed on the oscilloscope by *gently* touching one electrode to the heart and the other to body tissue. Cotton wick electrodes are best for this work. The sensitivity of the oscilloscope should be ~ 0.5 mV/cm. The effect of temperature changes on the heart can be investigated by using Ringer's solutions with $\sim 15°C \leqslant T \leqslant 35°C$.

Now cut through the frog's body just below the front legs. Cut away skin and muscle in the abdomen and clean out the cavity. Look for three silver cords running from each side of the spinal column toward each leg. These are the sciatic nerves. Clear away the membrane protecting each nerve. Slip forceps underneath the three cords, draw a black thread around them, and tie them off. Always handle the nerve by this thread. Cut the nerve free of the spinal cord and work down, so as to free it to the base of the leg. Keep the nerve fiber damp with Ringer's solution at all times. Then cut the skin from below the knee up to the pelvis and peel it back. Below the knee, you will see a fine silver thread; this is the other end of the nerve. Free it and tie it off with a thread. Then work back from this end until the entire nerve is free; try to get about 8 cm of fiber. Handle the nerve fiber by the threads and keep it moist at all times.

Transfer the sciatic nerve to the chamber. Begin by stimulating with 1 pulse/sec at an amplitude of 1 V and a pulse duration of 1 msec. Increase the amplitude voltage until the nerve signal spike reaches a maximum. The signal is a composite of the effects of several fibers because the sciatic nerve is a collection of fibers. The sciatic nerve has three different kinds of fibers, and each conducts at a somewhat different velocity. With care, the components can be observed because each conducts at a slightly different velocity and is triggered by a slightly different potential.

Lobsters (*H. americans*) are a convenient source for giant axons; the ventral nerve is composed of four of them. These can be obtained in the following way. Cut off one of the lobster's legs as close to the body as possible and bleed the lobster. Then remove the rest of the legs, both claws and the tail. Boil or broil the claws and tail and have them as a snack with half a bottle of white wine. Cut through the shell (on the dorsal side) on each side and bring the two cuts together just behind the eyes. Then, starting from the back, clean out the body cavity and cut away the muscle from the sides. When you get toward the front, be careful. The stomach is a gray sack just below the brain. Do not puncture it, but dissect around it. You'll see a cartilaginous channel along the middle of the ventral side. Start from the tail and cut along the channel. Grasp each side of the body and pull it apart to open the channel. You should now be able to see the nerve. Tie it and free it, working from the tail to the head. Transfer the nerve to the chamber and start with a stimulus voltage of ~ 0.1 V. This nerve is very sensitive to drying and may well function for only about an hour.

A useful bathing solution for the nerve fibers is:

1. 6.5 gm/liter NaCl

2. 0.14 gm/liter KCl

3. 0.12 gm/liter $CaCl_2$

4. 0.02 gm/liter $NaHCO_3$

5. 0.01 gm/liter NaH_2PO_4

6. 2.0 gm/liter glucose

In the next experiments, red blood cell ghosts are used to observe membrane transport, and the capacitance of the cell is measured in order to show that the membrane thickness is $\simeq 10$ nm. To prepare a sample of red blood cell ghosts, begin by obtaining a total of 100 ml of blood; somewhat outdated whole blood from a local hospital will be satisfactory. Immediately add 20 ml of 3.8% sodium citrate to prevent clot formation; then divide the sample among the required number of centrifuge tubes and spin them at 1600 g for 10 minutes. (Each centrifuge will have a chart showing the proper combination of revolutions per minute and rotor size.) Pipette off the supernatant and re-suspend the cells in the washing solution whose composition is given in the list of solutions at the end of these notes. Centrifuge again. Again remove supernatant, re-suspend, and spin. After that, re-suspend the final sediment in enough pH 7.4 buffer to give a volume of $\sim$ 25 to 30 ml.

The ghosts are now prepared by incubating a few milliliters of the final suspension for about 5 minutes at a temperature of $37°C$; the effects of different incubation temperatures may be investigated. Choose a hemolyzing solution and add about 30 to 35 ml; resume incubation at the previous temperature. Hemolysis will occur visibly. Wait for about 5 minutes and then add 1.75 ml of 3.32 M NaCl. If the ghosts were prepared at a temperature other than $37°C$, wait 5 minutes and then shift the ghosts to an incubator at $37°C$.

The ghosts are ready to "heal" and should be left at $37°C$ for about an hour. Examine them under the phase microscope after about 30 minutes to make sure the cells are normal. After about an hour, wash the ghosts by spinning them down twice at 34,500 g for 10 minutes. Then make a final suspension of ghosts in about 30 to 35 ml of medium. The ghosts should be kept at $37°C$ at all times.

The ghosts may be used for a variety of ion-transport experiments. These are illustrated in the flowchart given in Figure A1.5.

The electrical capacitance of a suspension of ghosts can be measured and the thickness of the blood cell membrane deduced from its capacitance. Place a cell suspension in a 120-ml chamber. The density of cells in the suspension is measured by counting the cells in a known volume with a microscope. The chamber is isolated in a constant temperature bath, and a mechanical stirrer is connected through the chamber

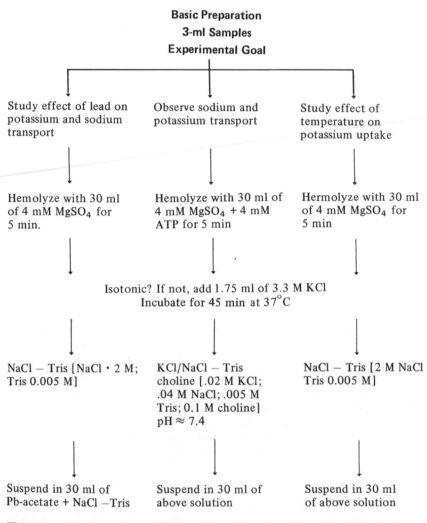

Basic Preparation
3-ml Samples
Experimental Goal

Study effect of lead on potassium and sodium transport

Observe sodium and potassium transport

Study effect of temperature on potassium uptake

Hemolyze with 30 ml of 4 mM MgSO$_4$ for 5 min.

Hemolyze with 30 ml of 4 mM MgSO$_4$ + 4 mM ATP for 5 min

Hermolyze with 30 ml of 4 mM MgSO$_4$ for 5 min

Isotonic? If not, add 1.75 ml of 3.3 M KCl
Incubate for 45 min at 37°C

NaCl − Tris [NaCl · 2 M; Tris 0.005 M]

KCl/NaCl − Tris choline [.02 M KCl; .04 M NaCl; .005 M Tris; 0.1 M choline] pH ≈ 7.4

NaCl − Tris [2 M NaCl Tris 0.005 M]

Suspend in 30 ml of Pb-acetate + NaCl −Tris

Suspend in 30 ml of above solution

Suspend in 30 ml of above solution

Then

Draw samples (~3 ml) at 2, 10, 20 40, and 80 min. Each sample goes into 30 ml of cold isotonic choline chloride (0.12 M) − lithium chloride (0.05 M) with 0.005 Tris adjusted to pH 7.4. Centrifuge at 35,000 g for 5 min. Extract pellet, dilute to 50 ml with H$_2$O and assay for Na and K.

Figure A1.5 The above chart summarizes some experimental procedures for studying ion transport across red blood cell ghost membranes.

by a flexible shaft so that the solution remains uniformly mixed. The electrodes, which are the plates of a capacitor, are then connected to a radio-frequency bridge. We have found it convenient to monitor the bridge circuit with a radio receiver; we also bring the output from the final amplifier through a 60-Hz rejection filter and a 10^3-Hz band-pass filter. Impedance measurements are then made as a function of frequency between 0.5 and 2.0 MHz. In this frequency range, difficulties with drift and electrode polarization are minimal. Capacitance curves for the suspension of ghosts and for the suspending solution alone are extrapolated to infinite frequency. The difference between the extrapolated values represents the capacitance introduced by the cells. Then, since

$$C_{\text{mem}} = \tfrac{1}{3} C_{\text{ghosts}} / n\pi r^4$$

where r is the radius of the cells, we have

$$\Delta x_{\text{mem}} = \frac{1}{4\pi(9 \cdot 10^{11})} \left(\frac{\epsilon}{C_{\text{mem}}} \right)$$

Since you should find

$$0.7 < C_{\text{mem}} < 1.3 \text{ F/cm}^2$$

and since $3 < \epsilon < 10$

$$\Delta x \approx 10 \text{ nm}$$

In our experience, confidence in the technique is gained if measurements on known impedances are first carried out; a very good test is to measure water as a function of temperature.

Proof that the capacitance measurements are actually connected with the cell membrane can be provided by adding a strong detergent to the suspension. The detergent breaks up the cell membrane very effectively, as examination under the microscope will show, and a sharp and easily observed change in the impedance of the suspension is observed.

In order to study artificial membranes of lipid bilayer, the experimental arrangement shown in Figure A1.6 should be obtained. The apparatus should be mounted on a vibration-free surface. The actual membrane is formed by touching a drop of lipid solution to the hole in the Teflon beaker with a thinned-out artists brush or a micro-syringe. Wipe the brush gently across the hole a few times. If all goes well,

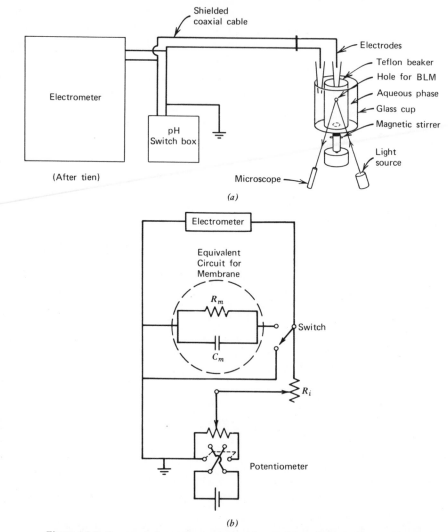

Electrometer

pH
Switch box

(After tien)

Electrodes

Teflon beaker

Hole for BLM

Aqueous phase

Glass cup

Magnetic stirrer

Light
source

Microscope

(a)

Electrometer

Equivalent
Circuit for
Membrane

R_m

C_m

Switch

R_i

Potentiometer

(b)

Figure A1.6 A suggested experimental setup for studying bilayer lipid membranes.

the drop will thin spontaneously and in reflected light will show first colored fringes, then a silver appearance, and finally a black surface. This last indicates a thickness of ~10 nm. If the membrane breaks, clean the brush with a 2:1 solution of chloroform and methanol and try again. Wipe the hole if repeated breaks occur.

To prepare the lipid solution, begin by washing all the glassware to be used with detergent and then with hot water and acetone. Then place 12 g of cholesterol in 300 ml of n-octane and bubble oxygen through the liquid at ~ 0.1 liter/min for ~ 6 hours while holding the liquid at its boiling point in a 1-liter boiling flask. Filter the solution until clear.

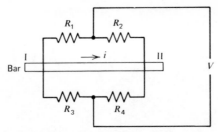

Figure A1.7 A schematic of a strain gauge arrangement for studying muscle behavior.

The resistance and capacitance of the membrane may be measured with the arrangement shown in Figure A1.6. The dielectric strength, which is the field strength at which the membrane breaks, may also be measured. More elaborate apparatus may also be employed so that the effects of various proteins on the membrane resistance can be studied.

The tension developed by a muscle under different conditions can be measured with a strain gauge. Four resistive elements are fixed to a metal bar and arranged as shown in Figure A1.7. The resistance of these elements is proportional to the force on them. Now if the bar (I II) bends, the force on two of the resistors increases and on the other two, it decreases. In equilibrium (no bending), no potential difference is observed between (I) and (II); when the bar is bent, current flows.

After first pithing a frog, remove the gastrocnemius muscle by opening the upper thigh and cutting the muscle free of the tendon; tie a string around it; then at the other end of the leg, cut through the femur. This will provide an attachment point for holding one end of the muscle. Keep the muscle wet with Ringer's solution.

Calibrate the gauge with a range of weights. Then tie the upper end of the muscle to the strain gauge attachment and balance the gauge with the external potential so that the weight of the muscle is taken out. Fix the femur end in a clamp and moisten the muscle with Ringer's solution.

Feed the output of the strain gauge to a chart recorder. You are now prepared to make several observations. By moving the femur portion, find the resting length of the muscle, which is the length at which a slight tension, P, is measured without any external stimulus. Stimulate the muscle with a voltage pulse and measure the tension. Shorten the muscle slightly (~ 2–3 mm) and repeat. Continue this until the tension is nearly zero. Then reverse the procedure until the muscle returns to the resting length. Continue increasing the length until the tension developed upon stimulation is nearly zero; then reverse the process. You can also investigate the effects the multiple stimuli and determine the speed of contraction and relaxation. Find the minimum value of the stimulus voltage that produces no contraction even if the stimulation is applied for a considerable time. By varying the times from this value, determine the relation between the strength, the contraction, and the duration of the stimulus.

app. 2

Notes

1. If light from a point source passes through an aperture, it will be diffracted by that aperture. If the aperture is a circle, the position of the first minimum in the diffraction pattern is

$$\theta = 1.22 \ \lambda/\text{diameter}$$

Now suppose we have a collimating lens; this lens serves to limit the beam. If we examine the image of a point source, we find it is a diffraction pattern of circular dark and light fringes, that is, the diffraction pattern of an aperture equal to the diameter of the collimating lens. Suppose there are now two point sources, separated by an angle α, as seen from the lens. If

$$\alpha = 1.22 \ \lambda /\text{lens diameter}, a$$

the center of one pattern is on the edge or central maximum of the other. Therefore, if two point sources are to be resolved.

$$\alpha > 1.22\lambda /a$$

2. The magnification of a microscope is

$$\text{Mag, } M = \left(\begin{array}{c} \text{enlargement produced} \\ \text{by the objective} \end{array} \right) \times \left(\begin{array}{c} \text{magnification} \\ \text{of the eyepiece} \end{array} \right)$$

From Figure A2.1,

$$M = (q/p) \times \left(\frac{25}{f} + \frac{25}{D} \right)$$

Now

$$f_e \ll D \qquad \text{so} \qquad 25/D \ll 25/f_e$$

$$p \approx f_0, \qquad \text{so we write} \qquad f_0 \text{ for } p$$

$$f_e \approx L \qquad \text{so we write} \qquad L \text{ for } q$$

whence

$$M = \frac{L}{f_0} \frac{25}{f_e}$$

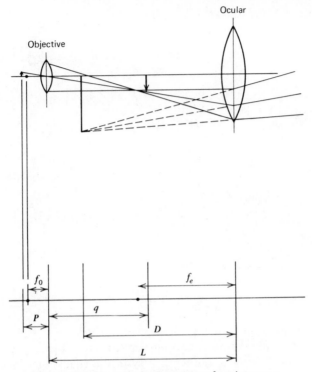

Objective

Ocular

f_0

P

q

D

f_e

L

Figure A2.1 The optical arrangement of a microscope.

3. Work is defined as

$$W \equiv \int \mathbf{F} \cdot d\mathbf{s} = m \int \frac{d\mathbf{v}}{dt} \cdot \mathbf{v} dt = \frac{m}{2} \int \frac{d}{dt}(v^2)dt$$
$$= \frac{m}{2}(v_2{}^2 - v_1{}^2)$$

where v_2 and v_1 are velocities of the mass m at time t_2 and t_1; $t_2 > t_1$.

If the net work done when the particle moves over a closed path is zero, then we can write

$$\oint \mathbf{F} \cdot d\mathbf{s} = 0$$

and

$$\nabla x \mathbf{F} = 0$$

However, the curl of any gradient is zero, so we can now write

$$\mathbf{F} = -\nabla(\text{scalar}) = -\nabla(\text{potential or potential energy})$$

Now

$$\overline{F} \cdot d\overline{s} = -\nabla V = -dV$$

or

$$F_s = \partial V/\partial s$$

and

$$W = V_1 - V_2$$

where the zero level of V is arbitrary.

4. Thermodynamics proceeds from two fundamental laws. The first is a statement of energy conservation, usually in the form: A system in equilibrium exchanges energy with the environment only by taking in heat, δQ, or doing work, dw. Thus

$$\delta Q = du + dw$$

where du is the internal energy of the system. The second law introduces a function, S, the entropy:

$$dS = dQ/T$$

If a system is isolated, no exchange of heat with the outside occurs. If a spontaneous process occurs irreversibly in such a system, the entropy always increases. From this, it follows that entropy is a measure of the degree to which the energy of a system has ceased to be available energy. The maximum entropy is therefore associated with the state of maximum disorder.

5. Consider the circuit in Figure A2.2. If we close the switch,

$$V_R = Ri = R dQ/dt$$

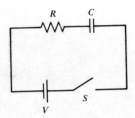

Figure A2.2 A standard RC circuit, as used in the derivation of the time constant.

and

$$V_c = Q/C$$

where Q is the amount of charge. So

$$V_E - R\frac{dQ}{dt} - \frac{Q}{C} = 0$$

Now the current at $t = 0$ is

$$i_0 = V_E/R$$

and

$$Q_{max} = CV_E$$

So

$$Q = CV_E(1 - e^{-t/RC})$$

or

$$i = (V_E/R)e^{-t/RC} = i_0 e^{-t/RC}$$

and

$$i \rightarrow i_0/e \qquad \text{when} \qquad t = RC$$

For this reason, the quantity RC is called the time constant of the circuit. Of course, the same result is obtained by analyzing the discharge of the capacitor from a fully charged condition.

problems

1. What is the volume of a unit cell with sides **a, b, c**?

2. The optical path difference between the incident and diffracted beam must be an integral multiple of the wavelength. Suppose you observe two beams scattered off a crystal; one satisfies the above condition but the other does not. Give an argument to show that the intensity ratio of the two is roughly equal to the number of unit cells in the crystal.

3. Is the density of atoms in a given plane related to the separation between planes? Support your answer.

4. Insulin has 4 molecules per unit cell; the sides of the unit cell are 44, 51.4, and 30.4 Å. Can the molecular weight of insulin be determined from this information?

5. Roughly how fast must a centrifuge rotor spin in order to give a molecule whose molecular weight is 10^5 Daltons, a gravitational potential energy equal to its random thermal energy? What is the effective gravity then?

6. What is the molecular weight of that macromolecule whose change in gravitational potential energy in moving 1 cm vertically equals kT?

7. Prove that at equilibrium $v_{max} = F/f$ and find the time needed to come to equilibrium (see Chapter 2).

8. Consider a centrifuge rotor $\omega = 20 \times 10^3$ r/min, containing tubes filled with 5 cm of a suspension of spherical macromolecules that are 100 nm in diameter. If the density of the macromolecules is ~ 1.3 and the rotor radius is 5 cm, find the least time needed to spin down the suspension.

9. The density of an *E. coli* cell is about 1.1. About how long would it take for *E. coli* to sediment 1 cm at 1 g?

10. For a certain macromolecule in water,

$$D = 6.3 \times 10^{-7} \text{ cm}^2/\text{sec}$$

If the sedimentation constant is 4.4×10^{-13} sec, where the specific volume is 0.75 cm^3/g and $\rho \approx 0.99$, what is the molecular weight of the macromolecule?

11. Consider an enzyme with $D = 8.2 \times 10^{-7}$ cm^2/sec. How long would it take for this enzyme to diffuse across an *E. coli* cell?

12. Consider a prolate macromolecule that has a molecular weight of 16,500 Daltons. If the specific volume is 0.75 cm^3/g and $D \approx 9.3 \times 10^{-7}$ cm^2/sec, show that the axial ratio of the macromolecule is about 10:1.

13. What is the interaction energy between a monovalent ion, say K or Na, and an H_2O molecule, **p** $= 6 \times 10^{-30}$ Cb/m for a separation of 1 nm?

14. What is the decrease in the ionic bond strength of NaCl (separation $\simeq 2.8$ Å) in water?

15. Consider the basic Michaelis-Menten scheme. Show that if the reaction is inhibited by the product, the plot of $1/v$ vs $1/[s]$ is a straight line.

16. Suppose one third of the protein in an *E. Coli* acts as an enzyme. If all protein in a daughter cell is new and the dividing time for *E. coli* is ~ 30 minutes, how many amino acids per sec^{-1} enzyme^{-1} are being produced?

17. Estimate the mass of the plasma membrane of *E. coli* if the molecular weight of a phospholipid molecule is $\sim 10^3$ Daltons. What is the mass if there is one protein, with a molecular weight of 25×10^3 per phospholipid?

18. The mobility ratio of Cl/Na is 8/5. Consider two dilute solutions of NaCl, one at 0.1 *M* and the other at 0.005 *M*. These are separated by a porous membrane. What is the potential drop across the membrane?

19. Consider a contractile vacuole in a particular protozoa. This vacuole, when filled with water, has a maximum diameter of about 7 μm and contracts to completely expel the water it accumulates. In order to keep the protozoan in equilibrium, the vacuole must contract about three times per minute. The volume of the protozoan is 16,800 μm^3, and its surface area is 3500 μm^2. From this data, due to Kitching, estimate the permeability of the surface membrane of the protozoan to water.

20. Use cable theory to get a first approximation for the velocity of the nerve impulse in the squid axon.

21. Show that the membrane relaxation time should be in the order of microseconds.

22. Prove that the time constant for a discharging capacitor is the same as for a changing capacitor.

23. Estimate the maximum hydrostatic pressure that a lipid bilayer can resist.

24. Consider a 10-ml sample of benzene irradiated at 280 nm for 10 minutes. If the temperature of the sample increases by $0.1°$, estimate the absorption coefficient.

25. H. Bull (*JACS*, 67, 4, 1954) gives the following for monolayers of egg albumin:

(mN/m) surface tension	0.07	0.11	0.18	0.20	0.26	0.33	0.38
(m²/mg) area/mg	2.00	1.64	1.50	1.45	0.38	1.36	1.32

What is the molecular weight and size of the albumin molecule?

26. Find the weight that is just slightly too heavy for a given muscle to lift.

27. Estimate the pressure increase required to maintain the same volume of blood flow in an artery whose diameter is reduced by 25% because of plaque formation.

28. The diameter of the aorta is about 2.5 cm; the average diameter of a capillary is about 8×10^{-4} cm ≈ 8 μm. There are probably about 4×10^9 capillaries in the circulatory system. If blood is an ideal fluid, estimate the velocity ratio of the blood in the aorta to the blood in a capillary, if the volume output of the ventricle is ~ 45 cm³ at 80 beats/min.

29. By what fraction of its diameter must an artery be reduced in order to have turbulent blood flow occur?

30. Consider a pair of capillaries branching from a small vein. If frictional minimization is the determining factor, show that half the angle between the capillaries should be given by

$$\cos^{-1} \Theta \simeq (r/R)^4$$

where r/R is the ratio of the radius of the capillary to the radius of the vein.

31. Show that the tymphanum moves $\approx .1$ nm or less in response to a sound wave that is in fact audible

32. Estimate the resolution of the eye. From this, estimate the number of rods/cm^2.

33. Estimate the breakdown voltage of a plasma membrane if the dielectric constant is ~ 6.

34. Estimate the net positive charge transferred when the nerve impulse goes from the resting potential to the peak.

35. For a squid axon, the leakage resistance R_m is ~ 2000 Ωcm^2, and the conductivity of the cytoplasm core R_i is ~ 200 Ωcm. First prove that the input impedance is given by

$$Z^2 = \frac{R_m R_i}{2\pi^2 \rho^3 \sqrt{1 + 4\pi^2 f^2 R_m{}^2 C_m{}^2}}$$

where ρ is the radius and f is the frequency appropriate to the value of Z. Then find the value of Z for squid axon.

36. Show that a nonmylenated fiber with a diameter of 5 μm is a leaky cable with $Z \approx 20 \times 10^6$ Ω.

37. Investigate how the frequency of visual signals due to thermal excitation depends on the energy threshold of rhodopsin.

38. In X-ray diffraction, would you expect the temperature of the crystal to affect the intensity of the reflection? Justify your answer.

39. Consider Na-transport in the gut. If $J_{m \to s}/J_{s \to m}$ is determined by passive ionic diffusion, predict the value of the ratio for potential differences of -3 mV, 0 mV, and $+7$ mV (see Chapter 11).

40. Do the above problem for Cl, where $\Delta V = -25$ mV, -3 mV, 0 mV, and $+7$ mV.

41. Suppose membranes were a double layer of protein. If the average energy of each is about five times the thermal energy, estimate the surface tension of the membrane for spherical proteins with diameters of 5 nm, packed so that the separation is about 10% of the diameter. How does this result compare with the surface tension of the lipid bilayer expressed as an energy per unit area?

42. *E. coli* must synthesize about 4×10^{-5} g of DNA per generation time. About how many nucleotides per second does this represent? What must be the rotational velocity of the double strand? Estimate that energy and compare it to the thermal energy.

bibliography

General Sources

Setlow, R. and Pollard, E.: *Molecular Biophysics*, Addison-Wesley, Reading, Mass., 1962.

Ackerman, E.: *Biophysical Science*, Prentice-Hall, Englewood Cliffs, N.J., 1962.

Patton, T. and Ruch, H.: *Physiology and Biophysics*, W. B. Saunders, Philadelphia, 1965.

Green, D.: *Molecular Insights into the Living Process*, Academic Press, New York, 1967.

Reviews of Modern Physics, January and April, 1959; pp. 1–563.
Lehninger, A.: *Biochemistry*, Worth, New York, 1975.

Chapter I

Barretal, G.: "Determination of the Masses of Viruses by Quantitative Electron Microscopy," *Q. Rev. Biophy.*, *9*, 459 (1976).

Bragg, L.: *The Development of X-ray Analysis*, Bell, London, 1975.

Beer, M.: "The Possibilities of Obtaining High Resolution (~ 30°A) Information Using the Electron Microscope," *Q. Rev. Biophy.*, *7*, 211 (1974).

Blake, C.: "Interpreting Protein Electron Density Maps," *Nature*, *262*, 1972 (1970).

Cosslet, V.: "High Voltage Electron Microscopy," *Q. Rev. Biophy.*, *2*, 95 (1969).

Dover, S.: "Models from Maps," *Nature*, *260*, 96 (1976).

Langer, R., et al: "Electron Microscopy of Thin Protein Crystal Sections, *J. Mol. Bio.*, *93*, 159 (1975).

Chapter II

Anderson, N.: "Preparative Particle Separation in Density Gradients," *Q. Rev. Biophy.*, *1*, 217 (1968).

Bickel, W., et al: "Application of Polarization Effects in Light Scattering: A New Biophysical Tool," *PNAS*, *73*, 486 (1976).

de Duve, C.: "Exploring Cells with a Centrifuge," *Science, 189*, 186 (1975).

Giglio, M., and A. Vendramini: "Soret-type Motion of Macromolecules in Solution," *Phys. Rev. Lett.*, *38*, 26 (1977).

Koch, A., and G. Blumberg: "Distribution of Bacteria in the Velocity Gradient Centrifuge," *Biophy. J.*, *16*, 389 (1976).

Rigler, R., and M. Ehrenberg: "Fluorescence Relaxation Spectroscopy in the Analysis of Macromolecular Structure and Motion," *Q. Rev. Biophy.*, *9*, 1 (1976).

Selser, J., et al: "A Light Scattering Measurement of Membrane Permeability," *Biophy. J.*, *16*, 1357 (1976).

Shindon, H., and J. Cohen: "Observation of Carboxyl Groups in Lysozyme by NMR," *PNAS*, *17*, 1979 (1976).

Yeh, Y., and R. Keeler: "A New Probe for Reaction Kinetics – The Spectrum of Scattered Light," *Q. Rev. Biophy.*, *2*, 315 (1969).

Chapter III

Allen, L.: "A Model for the Hydrogen Bond," *PNAS*, *72*, 4701 (1975).

Brant, D.: "Conformational Theory Applied to Polysaccharide Structure," *Q. Rev. Biophy.*, *9*, 527 (1976).

Cooper, A.: "Thermodynamic Fluctuations in Protein Molecules," *PNAS*, *73*, 2740 (1976).

Chothia, C.: "The Nature of the Accessible and Buried Surfaces in Proteins," *J. Mol. Bio.*, *105*, 1 (1976).

———: "Structural Invariants in Protein Folding," *Nature*, *254*, 304 (1975).

Creighton, T.: "The Homology of Protein Structures," *Nature*, *255*, 743 (1975).

Israelachvili, J.: "The Nature of Van der Waals Forces," *Contemp. Phys. 15*, 159 (1974).

———: "Van der Waals Forces in Biological Systems." *Q. Rev. Bioph.*, *6*, 341 (1973).

Levitt, M., and A. Warshal: "Computer Simulation of Protein Folding," *Nature*, *253*, 694 (1975); *256*, 238 (1975).

Levitt, M., and C. Chothia: "Structural Patterns in Globular Proteins," *Nature*, *261*, 552 (1976).

Nir, S., and M. Anderson: "Van der Waals Interactions Between Cell Surfaces," *J. Mem. Bio.*, *31*, 1 (1977).

Parsegian, V., and S. Brenner: "The Role of Long Range Forces in Ordered Arrays of Tobacco Mosaic Virus," *Nature*, *259*, 632 (1976).

Robson, B.: "Protein Folding Experiments," *Nature*, *262*, 447 (1976).

Chapter IV

Blake, C.: "X-ray Cryoenzymology," *Nature*, *263*, 273 (1976).

Cohen, R., and G. Benedek: "The Functional Relationship Between Polymerization and Catalytic Activity of GDH," *J. Mol. Bio.*, *108*, 151, 179 (1976).

Eigen, M.: "New Looks and Outlooks on Physical Enzymology," *Q. Rev. Biophy.*, *1*, 3 (1968).

Fink, A.: "Cryoenzymology: The Use of Sub-Zero Temperatures in the Study of Enzyme Mechanisms," *J. Theo. Bio.*, *61*, 419 (1976).

Perutz, M.: "X-ray Analysis, Structure and Function of Enzymes," *European J. Biochem.*, *8*, 455 (1969).

Robillard, G., et al: "Similarity of Crystal and Solution Structure of Yeast tRNA," *Nature*, *262*, 363 (1976).

Weber, M.: "The Role of Metal Ions in Carbonic Anlydrase," *J. Theo. Bio.*, *60*, 51 (1976).

Wooley, P.: "Models for Metal Ion Function in Carbonic Anlydrase," *Nature*, *258*, 677 (1975).

Koshland, D. "Protein Shape and Biological Control," *Sci. Am.*, October 1973.

Chapter V

Bolton, P., and D. Kearns: "NMR Evidence for Common Tertiary Structure Base Pairs in Yeast and *E. Coli* tRNA," *Nature*, *255*, 347 (1975).

Evans, F., and R. Sarma: "Nucleotide Rigidity," *Nature*, *263*, 567 (1976).

Hopfield, J., et al: "Proof-reading by Biological Molecules," *Phys. Today*, January 1977.

———: "Direct Experimental Evidence for Kinetic Proof-reading," *PNAS*, *73*, 1164 (1976).

———: "Proof-reading of the Codon-Anticodon Interaction on Ribosomes," *PNAS*, *74*, 198 (1977).

Parrish, J.: *Principles and Practice of Experiments with Nucleic Acids*, Wiley, New York, 1972.

Pullman, B.: "Quantum Mechanical Studies on Conformation of Nucleic Acids," *Nucleic Acid Research and Molecular Biology*, *18*, 1976.

Rodley, G., et al: "A Possible Conformation for Double Stranded Polynucleotides," *PNAS*, *73*, 2959 (1976).

Smith, T.: "Long Palindromes in Eucaryotic DNA," *Nature*, *262*, 255, (1976).

Weissman, M., et al: "Determination of Molecular Weights by Fluctuation Spectroscopy," *PNAS*, *73*, 2776 (1976).

Chapter VI

Blumenthal, R., B. Ginzburg, and A. Katchalsky: "Thermodynamic Model of Active Ion Transport in Erythrocytes," in *Hemorheology*, Pergamon Press, Oxford, 1967.

Bretscher, M.: "Directed Lipid Flow in Cell Membranes," *Nature, 260,* 21 (1976).

Carlemalm, E., and A. Wieslander: "Electron Diffraction Studies of Biological Membranes and Lipids," *Nature, 254,* 537 (1975).

Chung, S.: "Analysis of Membrane Noise," *Nature, 256,* 367 (1975).

Essig, A.: "Energetics of Active Transport," *Biophy. J., 15,* 651 (1975).

Fisher, K.: "Analysis of Membrane Halves: Cholesterol," *PNAS, 73,* 173 (1976).

Finean, J.: "Biophysical Contributions to Membrane Structure," *Q. Rev. Biophy., 2,* 1 (1969).

Fox, F.: *Membrane Molecular Biology*, Sinauer, Stamford, Conn., 1972.

Foster, K., et al: "The Electrical Resistivity of Cytoplasm," *Biophy. J., 16,* 991 (1976).

Issacson, L.: "Resolution of Parameters in Equivalent Electrical Circuit of Na Transport Mechanism," *J. Mem. Bio., 30,* 301 (1977).

Jaffe, L.: "Interpretation of Voltage – Concentration Relations," *J. Theo. Bio., 48;* 11 (1974).

Lang, M., et al: "Thermodynamic Analysis of Active Na Transport," *J. Mem. Bio., 31,* 19 (1977).

Luria, S.: "Colicins and the Energetics of Cell Membranes," *Sci. Am.,* December 1975.

Miller, D.: "Thermodynamics of Irreversible Processes," *Chem. Rev., 60,* 15 (1960).

Purvis, R.: "Microelectrodes in Spherical Cells," *J. Theo. Bio., 63,* 225 (1975).

Shields, R.: "Microtubules and Membrane Topography," *Nature, 256,* 257 (1975).

Scarborough, G.: "The Neurospora Plasma Membrane ATPase is an Electrogenic Pump," *PNAS, 73,* 1485 (1976).

Snell, F. (ed): *Physical Principles of Biological Membranes*, Gordon and Breach, New York, 1970.

Smith, W., and R. Mohler: "Necessary and Sufficient Conditions in the Tracer Determination of Compartmental System Order," *J. Theo. Bio., 57,* 1 (1976).

Ussing, H.: "The Interpretation of Tracer Fluxes in Terms of Membrane Structure," *Q. Rev. Biophy., 1,* 365 (1969).

Chapter VII

Adams, W.: "Upper and Lower Bounds on the Non-linearity of Summation of End Plate Potentials," *J. Theo. Bio.*, *63*, 217 (1976).

Anwyl, R., and P. Usherwood: "Ionic Permeability Changes Occurring at Excitatory Receptor Membranes of Chemical Synapses," *Nature*, *257*, 410 (1975).

Ardnt, R., and L. Roper: "A New Phenomenology for Squid Axon Voltage-Clamp Currents," *J. Theo. Bio.*, *48*, 373 (1976).

Adelman, W. (ed): *Biophysics and Physiology of Excitable Membranes*, Van Nostrand, New York, 1971.

Armstrong, C.: "Cellular Neurophysiology," *Q. Rev. Biophy.*, *7*, 179 (1974).

Bass, L.: "Current-Voltage Relations in Nerve Membranes," *J. Theo. Bio.*, *48*, 133 (1974).

Chung, S.: "Ineffective Nerve Terminals," *Nature*, *261*, 190 (1976).

———: "Analysis of Membrane Noise," *Nature*, *256*, 367 (1975).

Clay, J., and M. Shlesinger: "Theoretical Model of the Ionic Mechanism of $1/f$ Noise in Nerve Membrane," *Biophy. J.*, *16*, 137 (1976).

Cole, K.: "Electrical Properties of the Squid Axon Sheath," *Biophy. J.*, *16*, 137 (1976).

———: *Membranes, Ions and Impulses*, Univ. of California Press, Berkeley, 1968.

Colding-Jorgensen, M.: "A Description of Adaptation in Excitable Membranes," *J. Theo. Bio.*, *63*, 61 (1976).

Ehrenstein, G.: "Ion Channels in Nerve Membranes," *Physics Today*, October 1976.

Franks, N.: "Myelin Structure," *Nature*, *259*, 447 (1976).

Friesenstal, W.: "An Oscillating Neuronal Circuit Generating a Locomotory Rhythm," *PNAS*, *73*, 3734 (1976).

Hodgkin, A.: *The Conduction of the Nervous Impulse*, Charles C. Thomas, Springfield, Ill., 1964.

Martin, A.: "The Effect of Membrane Capacitance on Synaptic Potentials," *J. Theo. Bio.*, *59*, 179 (1976).

Neher, E., and B. Sakmann: "Voltage Dependence of Drug-Induced Conductance," *PNAS*, *72*, 2140 (1975).

Manresa, F., and H. Grundfest: "Temperature Dependence of the Four Ionic Processes of Spike Electrogensis," *PNAS*, *73*, 3554 (1976).

Sata, S., and N. Tsukahara: "Some Properties of the Theoretical Membrane Transients in Roll's Neuron Model," *J. Theo. Bio.*, *63*, 151 (1976).

Scott, A.: "The Electrophysics of a Nerve Fiber," *Rev. Mod. Phys.*, *47*, 487 (1975).

Swadlow, H., and S. Waxman: "Observations on Impulse Conduction Along Central Axons," *PNAS*, *72*, 5156 (1975).

Strandberg, M.: "Action Potential in the Giant Axon of Logigo," *J. Theo. Bio.*, *58*, 73 (1976).

Weisman, M.: "Models for 1/f Noise in Nerve Membranes," *Biophy. J.*, *16*, 1105 (1976).

Woolridge, D.: "Spike-forming Model of the Neural Membrane," *PNAS*, *72*, 3468 (1975); *73*, 2264 (1976).

Chapter VIII

Jones, M.: *Biological Interfaces*, Elsevier, Oxford, 1975.

Ohki, S., et al: "Monolayers at the Oil/Water Interface," *J. Theo. Bio.*, *62*, 389 (1976).

Torbet, J., and M. Wilkins: "X-ray Diffraction Studies of Lecithin Bilayers," *J. Theo. Bio.*, *62*, 447 (1976).

Tien, T.: *Bilayer Lipid Membranes*, Dekker, New York, 1974.

White, S.: "Formation of Planar Bilayer Membranes from Lipid Monolayers," *Biophy. J.*, *16*, 481 (1976).

Chapter IX

Bearden, A., and R. Malkin: "Primary Photochemical Reactions in Chloroplast Photosynthesis," *Q. Rev. Biophy.*, 7, 131 (1974).

Caplan, S.: "Nonequilibrium Thermodynamics and Its Application to Bioenergetics," in *Current Topics in Bioenergetics*, Academic Press, New York, 1971.

Dale, R., and J. Eisiger: "Intramolecular Energy Transfer and Molecular Conformation," *PNAS*, *73*, 271 (1976).

Fenna, R., and B. Mathews: "Chlorophyll Arrangement in a Bacteriochlorophyll Protein," *Nature*, *258*, 573 (1975).

Green, D.: "The Structure of Biological Membranes in Relation to the Principle of Energy Coupling," *J. Theo. Bio.*, *62*, 271 (1976).

Govinjee: "The Primary Events of Photosynthesis," *Sci. Am.*, December 1974.

Kemeny, G.: "Charge Pair – Model of Bioenergetics," *PNAS*, *73*, 2770 (1976).

Levine, R.: "The Mechanism of Photosynthesis," *Sci. Am.*, December 1969.

Lee, A.: "A Photosynthetic Structure," *Nature, 258,* 568 (1975).

Nachrransohn, D.: "Transduction of Chemical into Electrical Energy," *PNAS, 73,* 82 (1976).

Slater, E.: "The Coupling Between Energy-Yielding and Energy-Utilizing Reactions in Mitochondria," *Q. Rev. Biophy., 4,* 35 (1971).

Skulachev, V., et al: "Direct Measurement of Electric Current Generation by Cytochrome Oxidase," *Nature, 249,* 321 (1974).

Chapter X

Chapman, J., and E. Pollard: "Characteristics of Enzymatic Breakdown of DNA in Response to Ionizing Radiation," *Int. J. Rad. Bio., 15,* 323 (1969).

Coelho, A., et al: "Doubling of Fibroblasts from Different Species After Ionizing Radiation," *Nature, 261,* 586 (1976).

Cook, P., and I. Brazell: "Detection and Repair of Single-Strand Breaks in Nuclear DNA," *Nature, 263,* 679 (1976).

Dertinger, H., and H. Jung: "Molecular Radiation Biology," Springer-Verlag, New York, 1969.

Gray, L.: *Radiation Biology and Cancer* Pertner Foundation, Univ. of Texas, 1967.

Hildebrand, C., and E. Pollard: "The Study of Ionizing Radiation Effects of *E. Coli* by Density Gradient Sedimentation," *Biophy. J., 9,* 1312 (1969).

Hutchinson, F.: "The Lesions in DNA Containing 5-Bromouracil," *Q. Rev. Biophy., 6,* 139 (1973).

Neyman, J., and P. Puri: "A Structural Model of Radiation Effects in Living Cells," *PNAS, 73,* 3360 (1976).

Pollard, E.: "The Biological Action of Ionizing Radiation," *Am. Sci., 57,* 206 (1969).

Randolph, M.: "Degradation of DNA in *H. influenza* Cell After x-ray Irradiation," *J. Theo. Bio., 60,* 59 (1976).

Resnick, M.: "The Repair of Double-Strand Breaks in DNA," *J. Theo. Bio., 59,* 97 (1976).

Theus, R., et al: "Neutron Radiotherapy," *NRL Reports*, April 1973.

Biological Basis of Radiotherapy, published as *Brit. Med. Bull., 29* (January 1973).

Chapter XI

Curran, P., and S. Schultz: *Handbook of Physiology*, Am. Physio. Soc., Washington, D.C., 1968.

Curran, P., and J. McIntosh: "A Model System for Water Transport," *Nature, 193*, 347 (1962).

Hoshiko, T., and B. Lindley: "The Relationship of Ussing's Equation to the Thermodynamic Description of Membrane Permeability," *Biochem. and Biophys. Acta, 79*, 301 (1964).

Chapter XII

Chen, Y., and T. Hill: "Analysis of a Simple Prototypal Muscle," *PNAS, 71*, 1982 (1974).

Carlson, F.: "Structural Fluctuations in the Steady State of Muscle Contraction," *Biophy. J., 15*, 633 (1975).

Ebashi, S.: "Control of Muscle Contraction," *Q. Rev. Biophy., 2*, 351 (1969).

Goody, R., et al: "Cross Bridge Conformations as Revealed by X-ray Diffraction," *Biophy. J., 15*, 687 (1975).

Gray, B.: "Reversibility and Biological Machines," *Nature, 253*, 436 (1975).

Hill, T., and R. Simmons: "Free Energy Levels and Entropy Production in Muscle Contraction," *PNAS, 73*, 336, 2165 (1976).

Huxley, A., and R. Simmons: "Proposed Mechanism of Force Generated in Striated Muscle," *Nature, 233*, 533 (1971).

Mittenthal, J.: "A Sliding Filament Model for Skeletal Muscle," *J. Theo. Bio., 52*, 1 (1975).

Murray, J., and A. Weber: "The Cooperative Action of Muscle Protein," *Sci. Am.*, February 1974.

Squire, J.: "Muscle Filament Structure and Muscle Contraction," *Ann. Rev. Biophy. and Bioeng.*, 1975.

Sussman, M.: "Mechanochemical Availability," *Nature, 256*, 195 (1975).

Chapter XIII

Happner, D., and R. Plonsey: "Simulation of Electrical Interaction of Cardiac Cells," *Biophy. J., 10*, 1057 (1970).

Miller, C., et al: "A Theoretical Evaluation of Cardiac Output," *J. Theo. Bio., 63*, 89 (1976).

Sachs, F.: "Electrophysiological Properties of Tissue Cultured Heart Cells," *J. Mem. Bio.*, *28*, 373 (1976).

Woodcock, J.: "Physical Properties of Blood and Their Influence on Blood Flow Measurement," *Rept. on Prog. Phys.*, *39*, 1976.

Chapter XIV

Bekesey, G.: *Experiments in Hearing*, McGraw-Hill, New York, 1960.

Johnstone, B., and P. Sellick: "The Peripheral Auditory Apparatus," *Q. Rev. Biophy.*, *5*, 1 (1972).

Moller, A.: "Coding of Sounds in Lower Levels of the Auditory System," *Q. Rev. Biophy.*, *5*, 59 (1972).

Robertson, D.: "Correspondence Between Sharp Tuning and Two-Tone Inhibition in Primary Auditory Neurones," *Nature*, *259*, 477 (1976).

Weaver, E., and C. Gans: "The Caecilian Ear," *J. Expt. Bio.*, *191*, 63 (1975) and *PNAS*, *73*, 3744 (1976).

Chapter XV

Ashmore, J., and G. Falk: "Absolute Sensitivity of Rod Bipolar Cells in a Dark Adapted Retina," *Nature*, 263, 248 (1976).

Barlow, H.: "Retinal Noise and Absolute Threshold," *JOSA*, *46*, 634 (1956).

Bouman, M.: "My Image of the Retina," *Q. Rev. Biophy.*, *2*, 25 (1969).

Downer, N., and S. Englander: "Molecular Structure of Membrane Bound Rhodopsin," *Nature*, *254*, 625 (1975).

Ebray, T., and B. Honig: "Molecular Aspects of Photoreceptor Function," *Q. Rev. Bioph.*, *8*, 129 (1975).

Fein, A., and J. Charlton: "Local Membrane Current in Limulus Photoreceptors," *Nature*, *258*, 250 (1975).

Levine, D., and S. Grossberg: "Visual Illusions in Neural Networks," *J. Theo. Bio.*, *61*, 477 (1976).

Muijser, H.: "Photopigment Conversions Expressed in Receptor Potential and Membrane Resistance," *Nature*, *254*, 520 (1975).

Rushton, W.: "Pigments and Signals in Color Vision," *J. Physio.*, *220*, 1 (1972).

———: "Visual Pigments and Color Blindness," *Sci. Am.*, March 1975.

Saibil, H., et al: "Neutron Diffraction Studies of Retinal Rod Outer Segment Membranes," *Nature*, *262*, 266 (1976).

Werblin, F.: "The Control of Sensitivity in the Retina," *Sci. Am.*, January 1973.

Young, R.: "Visual Cells," *Sci. Am.*, October 1970.

Chapter XVI

Beland, P., and D. Russell: "Biotic Extinctions by Solar Flares," *Nature, 263,* 259 (1976).

Eigen, M.: The Origin of Biological Information in *The Physicist's Conception of Nature*, Reidel, Boston, 1973.

Hoyle, F., and N. Wickramasighe: "Primitive Grain Lumps and Organic Compounds in Carbonaceous Chondrites," *Nature, 264,* 45 (1976).

Kovacs, K., and A. Garay: "Primordial Origins of Chirality," *Nature, 254,* 538 (1975).

Latham, J.: "Possible Mechanisms of Corona Discharge Involved in Biogenesis," *Nature, 256,* 34 (1975).

Osterberg, R.: "Origin of Metal Ions," *Nature, 249,* 382 (1974).

Prigogine, I., et al: "Thermodynamics of Evolution," *Physics Today*, November and December 1972.

Smith, T.: "The Origins of Nuclei and Eucaryotic Cells," *Nature, 256,* 463 (1975).

Wong, J.: "The Evolution of a Universal Genetic Code," *PNAS, 73,* 2336 (1976).

Q. Rev. Biophy, 4, 77–276 (1971).

Chapter XVII

Cho, Z. (ed): "Physical and Computational Aspects of 3-D Image Reconstruction," *IEEE*, NS-21, June 1974.

Cohen, D.: "Magnetic Fields of the Human Body," *Physics Today*, August 1975.

Donald, I., and S. Levi: *Present and Future of Diagnostic Ultrasound*, Wiley, New York, 1976.

Gordon, R., et al: "Image Reconstruction from Projections," *Sci. Am.*, October 1975.

Gutch, C.: "Artificial Kidneys: Problems and Approaches," *Annual Reviews of Biophysics and Bioengineering*, Palo Alto, 1975.

Hussey, M.: *Diagnostic Ultrasound*, Wiley, New York, 1975.

Sauer, F.: "Non-equilibrium Thermodynamics of Kidney Tubule Transport," *Hdbk. of Physiology: Renal Physiology*, Am. Physiol. Soc., Washington, D.C., 1973.

Ullrick, K.: "Permeability Characteristics of Mammalian Nephron," *Hdbk. of Physiology*, Am. Physiol. Soc., Washington, D.C., 1973.

index